MOVING BOUNDARY PROBLEMS IN HEAT FLOW AND DIFFUSION

MOVING BOUNDARY PROBLEMS IN HEAT FLOW AND DIFFUSION

BEING THE PROCEEDINGS OF THE CONFERENCE HELD AT THE UNIVERSITY OF OXFORD 25–27 MARCH 1974

EDITED BY

J. R. OCKENDON, AFIMA
UNIVERSITY OF OXFORD

AND

W. R. HODGKINS, AFIMA
ELECTRICITY COUNCIL RESEARCH CENTRE, CAPENHURST

CLARENDON PRESS · OXFORD
1975

Oxford University Press, Ely House, London W.1

GLASGOW NEW YORK TORONTO MELBOURNE WELLINGTON
CAPE TOWN IBADAN NAIROBI DAR ES SALAAM LUSAKA ADDIS ABABA
DELHI BOMBAY CALCUTTA MADRAS KARACHI LAHORE DACCA
KUALA LUMPUR SINGAPORE HONG KONG TOKYO

ISBN 0 19 853345 4

PRINTED IN GREAT BRITAIN BY
J. W. ARROWSMITH LTD., BRISTOL

PREFACE

When the proposal to hold this conference was first put to the Institute of Mathematics and its Applications by the Oxford Study Group with Industry, it reflected their experience that research workers in a wide range of areas were attacking similar problems in heat flow and diffusion involving phase changes or chemical reaction and absorption. Such problems could roughly be classified as generalised Stefan problems and lay outside the scope of the standard texts. The wide range of application areas had brought about a situation where the results in one field were not readily available to the researcher in a different field.

The primary aim of the conference was thus to bring together these research workers from as many diverse fields of study as possible along with interested applied mathematicians and numerical analysts. Moreover, since only one text book and a handful of review articles on Stefan problems existed in the literature, it was decided to publish the proceedings.

For the purposes of this book the papers have been arranged in three parts. Part I deals primarily with applications and comprises papers which either survey a field of interest, say within a particular industry, or deal with the solution of specific problems. Part II consists of theoretical papers which seek to classify the problems and suggest techniques of solution, one dimensional problems receiving closest attention. The discussion covers both analytical and numerical techniques which frequently complement each other in a very satisfactory way. Both parts I and II emphasise problems where nonlinearity enters primarily through the presence of a moving boundary and effects such as convection are either neglected or treated in a simple, often empirical way. The considerable extra complexity arising from the presence of fluid flow is apparent in Part III which deals with thermal explosions. The papers in this part discuss mechanisms for the break-up and mixing of fluids with different heat contents which may give rise to explosive effects.

Items contributed to the discussion sessions at the conference have been placed at the end of the appropriate parts.

The editors have tried to make the book as accessible as possible by the use of cross-referencing both within and between the different parts. In the course of carrying this out, it became apparent that some of the results in Part II require only minor modification to shed light on problems in Parts I and III and equally several of these problems suggest ways in which the theoretical techniques themselves may usefully be generalised. It was also clear that while several of the mathematical techniques appear to have a fairly general range of applicability the number of methods used in the applications is almost as great as the number of problems. Each new physical effect apparently necessitates mathematical devices which must be incorporated in an *ad hoc* manner. Nonetheless, we feel that for conventional Stefan problems involving heat conduction or diffusion, there is now available a sound theoretical basis which enables a number of numerical schemes to be used with reasonable assurance although further refinements and comparison are required. However, in situations involving effects such as combined heat and mass transfer or abrupt changes in the physical parameters, there is a lack of analytical and numerical theory but even these difficulties seem insignificant compared to those encountered in convective problems.

The editors would like to express their thanks to the chairman of the conference, Mr. R.A. Scriven (Hon. Sec. IMA), for his encouragement and helpful suggestions concerning its organisation. We should also like to thank Mrs. Laurel Middleton and Mrs. Maureen Downie for typing the manuscript and preparing the figures.

CONTENTS

Part 1. Applications

Introduction

Part I of the proceedings is devoted to practical applications and consists of expository papers covering particular scientific or industrial areas together with shorter contributions providing a more detailed examination of specific aspects of these areas.

Historically, Stefan type problems have probably been of most industrial importance in the study of the solidification of metals and the first paper by Peel surveys some of the main areas in the steel industry where moving boundaries must be considered as an integral part of the problem. Amongst these is the problem of scrap melting considered in more detail by Perkins. This emphasises that whilst numerical techniques are just about adequate to tackle the problem of heat conduction with a moving boundary, the modelling of the boundary in the liquid region is still inadequate and empirical fitting of a heat transfer coefficient is needed for good agreement with experiment. The problem of incorporating carbon diffusion, in practice a vitally important element in determining the melting temperature of iron, is an area where lack of accurate data requires empirical fitting of results. The paper by Hodgkins and Waddington on metal solidification in furnace valves also highlights the observation that the results depend critically on the assumptions made about heat transfer in the liquid phase and across the interfacial boundary.

The paper by Andrews and Atthey investigating the ablation of metals using a laser beam considers both analytical and numerical techniques for a one-dimensional model of the process. An implicit integration technique is used for the numerical method. Within the limitations of the model the techniques appear satisfactory, although in practice the effects of melting at the edges and the outward conduction of heat may be significant.

Longworth considers the welding of two materials heated by the passage of an electric current where the electrodes are moving relative to the material. A numerical formulation is used in which the primary variable is heat content, rather than temperature; no difficulty occurs from incorporating Joule heating.

Besides the simple problems of melting and ablation which only involve the transfer of heat, many problems also involve the diffusion of other substances. Reference has been made to the rate-determining effect of the diffusion of carbon in iron melting and Peel refers to corrosion and oxidation problems where similar transfer mechanisms are involved. He also deals with the problem of determining the structure of the solid from a knowledge of the rate of cooling and of the chemical composition of the melt. Crank in his survey paper deals mainly with diffusion in chemical and biological problems. The problems are treated as one-dimensional and complexities in the formulation arise from lack of knowledge of diffusion rates and the relations between the behaviour of the material and the varying parameters. These are often nonlinear and cause further difficulties in the solution of the problem. In the one-dimensional case, expansions for short times and asymptotic approximations for long times are often possible whilst the Goodman heat-balance method is also useful for some problems.

Another industrial field involving moving boundary problems is that of glass manufacture which is surveyed in the paper by Gelder. He concentrates attention first on batch melting where a one-dimensional model using an implicit numerical integration method provides sufficient accuracy for the model chosen, and secondly on the problem of thermal switching of glasses in which the electrical properties of certain glasses change by several orders of magnitude on increasing the temperature by Joule heating.

The problem of the freezing of intrusions within the earth considered by Turcotte, although orders of magnitude larger than the previous applications, gives rise to similar problems. Here the scale is such that the convective heat transfer within the molten intrusion is of importance and the stability of the convective flow requires examination. The structure of the solidified phase provides experimental evidence of the temperature gradients which must exist within such flows.

On an even grander scale the phenomena of the evolution of stars considered by Eggleton provide problems involving moving boundaries. Material properties can change drastically throughout the interior of a star and careful and flexible division of the regions of the star is required to yield a tractable numerical problem. The thermal instability which causes the appearance and disappearance of convective regions within the star is found to be important in governing the course of stellar evolution. An implicit numerical scheme of integration is used for the spherically symmetric problem but little idea of accuracy is possible and the reliability of the method is assessed more by the qualitative predictions of the model.

The contributed note by Hebditch provides some experimental data on the solidification process in an Pb-Sn alloy at a phase boundary where both mass and heat transfer are important; it shows how, even in a simple case, accurate treatment of solidification may require a model far more complex than those used hitherto if accurate predictions of other than gross effects are required.

In general the application papers emphasise several areas where more comprehensive analytical and numerical techniques would be of value. Amongst these are the modelling of the solidification process, especially where mass and heat convection are important, the development of simple analytic ideas to encompass problems where length and time scales vary enormously, and the provision of simple techniques and criteria which might enable a wide range of problems to be classified and summarised, such as has occurred for classical heat conduction and diffusion problems. We consider these features further in Part II.

SOME MOVING BOUNDARY PROBLEMS IN THE STEEL INDUSTRY

D.A. Peel
(British Steel Corporation)

1. INTRODUCTION

Unlike the academic, the mathematician in industry is seldom able to choose the type of problem that he works on, much less to specialise on a particular topic. Since moving boundary problems only form a small proportion of the total, any individual industrial mathematician is unlikely to meet many of them at any one time. However, within something as large and as complex as the steel industry, it is not surprising to find that many of these problems exist and also that for quite a few of them little significant progress towards usable solutions has yet been made.

In attempting to select a few representative moving boundary problems to describe in detail, a classification scheme has been devised based upon the morphology of the moving boundary. Further classification has been made of the types of model that can be built in attempting to solve any particular problem and thus the kind of solutions that can reasonably be expected from any class of model. By solutions we mean rather more than the strict mathematical answers, either in the form of closed expressions or as numerical algorithms. Any model that is built is primarily for the purpose of answering specific questions, but the better a model is, in some sense, the more insight it can give in areas away from its principal target. In general, this is achieved by having more physics in it and less statistics.

In the Industrial Research field, a good model is one which generates unexpected but confirmable results and in so doing triggers off fresh directions of research, particularly on the experimental front.

And so, we can define three broad classes of models:

Class *C* models rely heavily on empirical relationships and *ad hoc* scaling factors. These models are perfectly suitable for examining existing plant design or operating practice. Class *C* for crude.

Class *B* models, *B* for better, or perhaps even best, are fairly free from empiricism, relying almost entirely on the laws of physics and chemistry and hardly at all upon the fruits of the statistician's art. Models of this type are essential in those situations where it is necessary to gain a deeper understanding of a process before designing a plant or operating it in a manner significantly different from usual and these models frequently have this desirable bonus of triggering off fresh lines of research. It is suspected that Class *B* models are created when one attempts to formulate a model precisely, without any concessions to the method of solution.

Class *A* models fall into two sub-classes. *A1* models are those few where an otherwise Class *B* model can be formulated and for which neat elegant solutions of wide ranging applicability can be found, *e.g.*, Fourier's heat conduction equation, as evidenced by Carslaw and Jaeger [2]. *A2* models, however, result when, in an attempt to achieve the elegance, reality is sacrificed by being cavalier with the assumptions. An obvious example is where linearisation is imposed, thus limiting the range of applicability of the model. There is of course no harm in this while the restrictions are borne in mind. But what is not generally realised is that the digital computer, whose advent has so expanded our ability to create and use complex models, cannot distinguish whether any particular user has regard for or understanding of such restrictions. This class of model is thus potentially dangerous, especially in industry.

2. PSEUDO BOUNDARIES

Problems are often stated in the form "how does the

position of that observable boundary change with time," but sometimes, after the situation has been analysed and the model built and validated, it turns out that the question has been answered without at any time invoking an actual moving boundary. Usually the boundary turns out to be just a sharp gradient in one or more of the variables but this is not always the case. It is likely that when this happens we have a Class *C* solution masking the need for a Class *B* model of a deeper level problem.

As an example, consider the phenomenon of decarburisation. Metallurgists are very fond of cutting open pieces of metal which have been subjected to some treatment, etching them by dipping the bits in a witch's brew and squinting at the results under a microscope. They frequently see a distinct boundary separating regions with different structure and want to relate the position of this boundary to the duration or intensity of the treatment. Decarburisation of the surface of a billet occurs within a reheating furnace whilst the billet is being prepared for further rolling operations. To the metallurgist it shows up as a particularly clear boundary just under the surface of the metal. It arises because carbon within the steel oxidises at the surface, setting up a concentration gradient so that more carbon diffuses to the surface, to be oxidised in its turn. This is particularly severe above a critical temperature (910°C). The resulting surface layers are of course no longer the same type of steel as the rest of the billet and frequently have to be ground off. Since the depth of metal affected can be of the order of several millimetres, the cost is considerable. The problem is even more serious if plant breakdowns occur immediately downstream of the furnace, since the metal may then remain at an elevated temperature for longer than planned.

The reverse situation is, however, put to good use in the operation of case hardening where a finished part is given a hard skin by holding it at a high temperature for an extended period, surrounded by graphite. Sufficient carbon diffuses in to enhance the surface properties and yet leave the core in its original tough state.

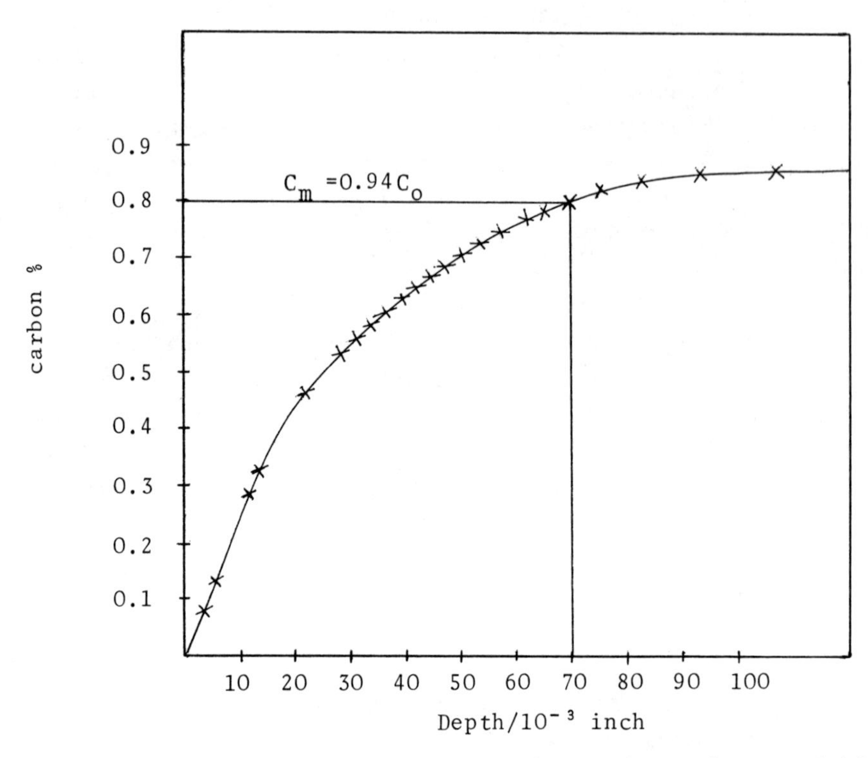

Fig. 1. Carbon concentration near the surface of a steel billet (redrawn from [1]).

Some results obtained by Birks and Nicholson [1] are reproduced in Fig. 1. They have shown that a simple diffusion model explains the measured carbon distribution near the surface but that the position of the change in structure seen by the metallurgist correlates very well with the position of the isopleth representing 94% of the initial carbon content. It may be that a deeper model concerned with the etching behaviour of individual metal crystals would explain the significance of the carbon level at which such sharp differences are observed.

Leaving aside pseudo-moving boundary problems of this type we can divide the remaining problems into those where the boundary is sharp and those where it is extended into a moving region.

3. SHARP BOUNDARIES

Any number of truly sharp moving boundary problems exist in one single field, that of corrosion. The simplest is that of scale formation. Like decarburisation, this occurs at a much faster rate above a critical temperature (570°C) and so can lead to considerable loss of metal within reheating furnaces. Losses of 3% are not unusual and 10% is not unknown. Again, these losses can be aggravated by plant breakdowns.

A considerable literature exists on the subject of predicting scale losses but it is all to do with the merits of fitting parabolic, logarithmic or other simple mathematical curves to experimental data (Class *C* solutions). No use appears to have been made of the obvious model relating the diffusion of oxygen through the scale with the boundary condition that the oxygen flux at the metal surface should equal the stoichiometric requirement for the oxidation of the metal at the rate the boundary moves (Fig. 2a). And yet Crank [3] gave virtually this form almost twenty years ago, under the heading, "Tarnishing Reactions". In practice, the situation with iron (or steel) is more complicated, since at these elevated temperatures, a succession of layers of different oxides builds up (Fig. 2b). A model to cope with this is conceptually no more difficult but would be very difficult to validate since the required diffusion coefficients would have to be known with considerable precision. Moreover, one school of thought holds that the scaling proceeds not by oxygen diffusing through the scale to the metal surface but by iron diffusing outwards until it meets the air. In the single layer case, the mathematical formulation of this yields identical equations, although the diffusion coefficients are different and the stoichiometric coefficients are reciprocal to each other. The use of this boundary condition should thus resolve the controversy.

Whichever mechanism is responsible for scaling, it is clear that it can be greatly affected by the composition of the furnace atmosphere and that these atmospheres are changing nowadays due to the changeover from Middle East oil to North Sea gas. The time is thus right for the emergence of Class *B* models in this area.

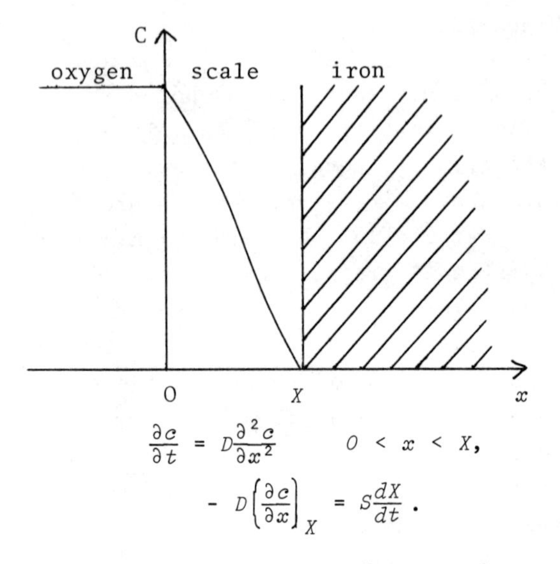

$$\frac{\partial c}{\partial t} = D\frac{\partial^2 c}{\partial x^2} \qquad 0 < x < X,$$

$$-D\left(\frac{\partial c}{\partial x}\right)_X = S\frac{dX}{dt}.$$

Fig. 2a. Simple model of scale formation.

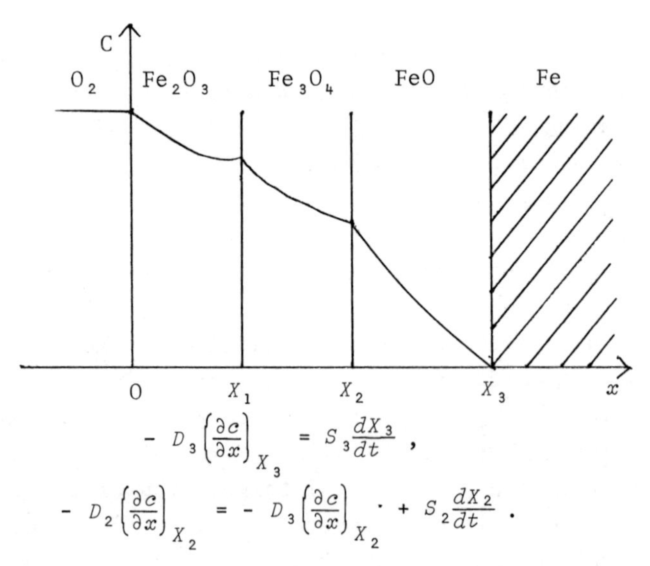

$$-D_3\left(\frac{\partial c}{\partial x}\right)_{X_3} = S_3\frac{dX_3}{dt},$$

$$-D_2\left(\frac{\partial c}{\partial x}\right)_{X_2} = -D_3\left(\frac{\partial c}{\partial x}\right)_{X_2} + S_2\frac{dX_2}{dt}.$$

Fig. 2b. Multi-layer scale formation.

4. EXTENDED BOUNDARIES

Examples of extended moving boundaries, or moving regions, are very common in the course of solidification processes. Pure substances, like water, freeze with a sharp boundary but mixtures, alloys and commercially pure substances *i.e.* most engineering materials, undergo some form of segregation on freezing which can give rise to an intermediate mushy zone between the liquid and the solid.

Industrially, solidification is an extremely important process.* All steel is solidified at least once and the same is true of other metals and many other materials. The problems that arise are of two sorts, engineering and metallurgical.

On the engineering side, the problems are concerned with designing and operating plant to remove heat in as efficient a manner as possible. Generally speaking, Class *C* models are adequate; the heat conduction equation is augmented by some means of accounting for the latent heat evolved during solidification. By defining two temperatures, called the liquidus and the solidus, the temperatures at which solidification begins and ends and by assuming some pattern of latent heat release between these two limits, one can either include an extra term in the heat conduction equation or modify the specific heat. This approach works quite well but it has not made use of any moving boundary notions; the mushy region is simply delineated by the liquidus and solidus isotherms. Nevertheless, very important practical problems have been tackled this way. For instance, in a continuous casting plant it is important not to cut the cast strand up into blooms or slabs until after the centre has finally solidified. Any liquid exposed at that stage would have 40 to 50 feet head of liquid steel behind it and would present a moving boundary of an altogether more spectacular kind!

However, this approach can only succeed in situations where the liquidus and solidus are known fairly accurately but, as we shall see below, these temperatures and particularly the solidus, vary quite a lot with the speed with which solidification proceeds. It is thus not possible to make a model of this type of, say, a continuous casting process in which the casting speed is allowed to vary over any large range without taking this into account in some way or another.

---

*See Hebditch (p. 112).

The metallurgical problems which arise are also concerned with an ability to calculate solidus temperatures. Steel is not just one material: it is a whole range of related materials. Clearly one would expect differing properties from alloys of different compositions but even limiting oneself to steel of one particular analysis, it is possible to achieve an enormous range of mechanical properties by means of different fabrication techniques and by heat treatment. This is because these processes modify the structure of the metal that is set up when it is cast. An ability to tailor this initial cast structure would extend the range of properties obtainable or make some particular properties obtainable with less effort. But before we can do this we must gain a much deeper quantitative understanding of the way this structure arises.

The discussion here is limited to dendritic freezing of a eutectic mixture; it is this type that gives rise to a mushy zone, although the principles involved can be applied equally well to other forms of freezing. This mushy zone occurs when the rate of solidification is not too great, which is of course quite common with industrially sized lumps of material which cool quite slowly. Under these conditions, a plane freezing interface becomes unstable and throws out long fingers, the primary dendrites, which subsequently throw out their own side fingers, the secondary dendrite arms, which in turn throw out tertiary arms (Fig. 3). By this time, remembering that it is happening in three dimensions, a considerable volume of the material is divided up into small cells with appreciable resistance to the flow of liquid between them. These cells continue to solidify as small individual pools. Small, in this context, means in the range 10 to 100 microns.

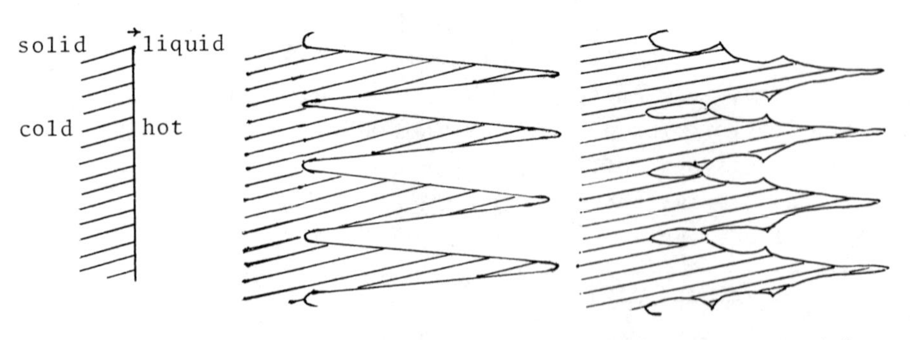

(a) Plane solidification (b) Primary dendrites (c) Secondary dendrites

Fig. 3. Formation of cells in dendritic freezing.

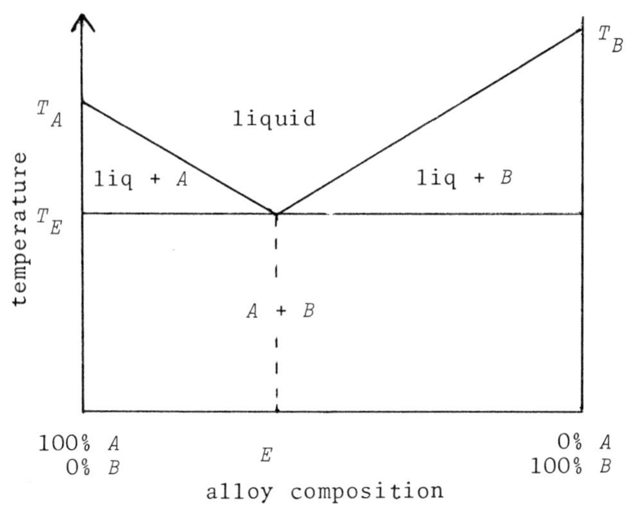

Fig. 4. Equilibrium diagram for binary alloy.

When a mixture freezes very, very slowly under what are called equilibrium conditions, it does so at a temperature which is a function of its composition, (*cf.* the depression of the freezing point of water when salt is added to it). The same thing can happen with metals as shown in Fig. 4. Metal A freezes at T_A and metal B at T_B, but some intermediate alloys freeze at temperatures lower than either of these. The minimum temperature T_E is known as the eutectic temperature and E denotes the eutectic composition.

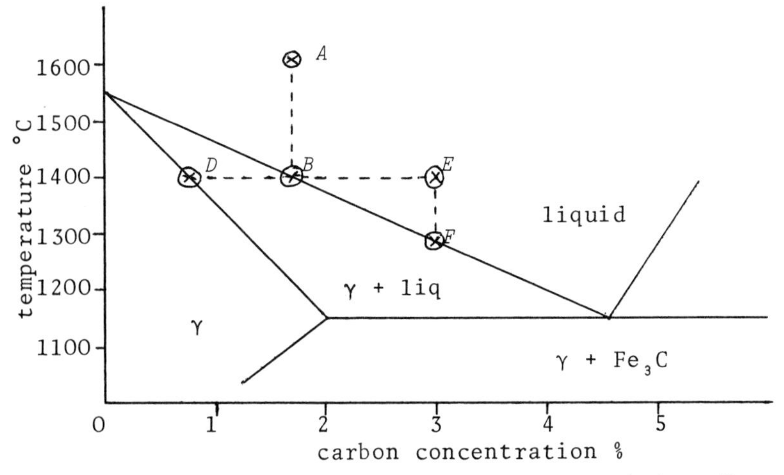

Fig. 5. Part of iron-carbon equilibrium diagram.

Fig. 5 shows one end of the equilibrium diagram for the iron-carbon system, rather simplified, which shows the regions in which the various phases are stable. From it several important points can be deduced. Liquid metal at the point A (1.6% carbon, 1600°C) cools without any solidification until it reaches B when its temperature has fallen to 1400°C. The first metal to solidify at this temperature will have a carbon content given by the position of D (.7% carbon) and will thus reject some of its carbon ahead of the liquid/solid interface. The ratio of the carbon contents of the liquid and the solid which are in equilibrium with each other is known as the partition coefficient.*

This rejected carbon consequently piles up at the liquid/solid interface thus raising the local carbon concentration above the initial level elsewhere in the liquid *e.g.* to the point E. At this concentration however, more freezing could not take place until the temperature fell back to the liquidus line at point F. But of course, before this happens, the local peak of carbon can begin to diffuse away into the liquid so that the point E moves back towards B. In practice a balance is maintained in which the rate at which the boundary moves determines the rate at which carbon is rejected and this must equal the diffusion flux away from the boundary. The situation at some later time is as shown in Fig. 6. The dimension η represents half the average spacing between adjacent secondary dendrite arms and is thus quite small.

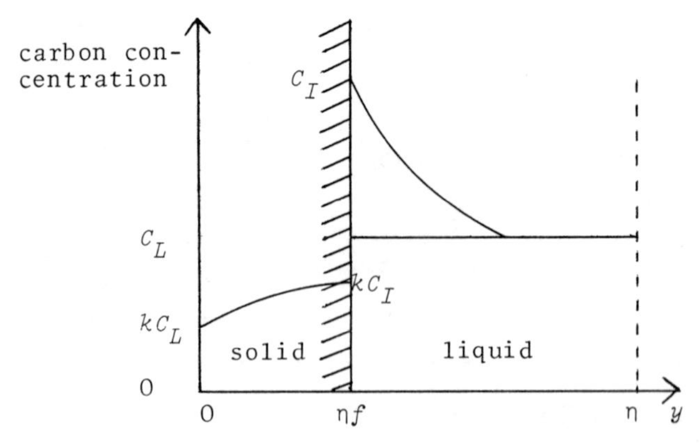

Fig. 6. Carbon concentration within a partly frozen cell

* See Tayler (p. 132).

We can now set up various models* of this dendritic freezing of a binary eutectoid alloy which will enable us to determine the distribution of solute across the individual cells. This distribution largely defines the metallurgical structure which is the starting point for the modifications which will be imposed by fabrication and heat treatment. Consider first transient freezing in one space dimension. If the half length of the rod is x_1, then

$$\frac{\partial}{\partial x}\left\{K\,\frac{\partial u}{\partial x}\right\} = \rho c\,\frac{\partial u}{\partial t} - L\rho\,\frac{\partial f}{\partial t}\,, \qquad 0 < x < x_1 \tag{4.1}$$

where the last term represents the latent heat evolution, $f(x,t)$ being the fraction solid. A variety of initial and boundary conditions could be used, for example,

$$u = u_1, \qquad t = 0 \tag{4.2}$$

$$u = u_2, \qquad x = 0, \qquad t > 0 \tag{4.3}$$

$$\frac{\partial u}{\partial x} = 0\,, \qquad x = x_1, \qquad t > 0. \tag{4.4}$$

Within a cell at a point x along the rod

$$\frac{\partial c}{\partial t} = D\frac{\partial^2 c}{\partial y^2}\,, \qquad \eta f < y < \eta \tag{4.5}$$

where

$$c = C_L, \qquad t = 0 \tag{4.6}$$

and

$$\frac{\partial c}{\partial y} = 0\,, \qquad y = \eta, \qquad t > 0, \tag{4.7}$$

C_L being the initial carbon content of the liquid metal. Since the freezing interface is in equilibrium

$$c = \frac{u_0 - u}{a}\,, \qquad y = \eta f \tag{4.8}$$

where u_0 is the melting point of pure solvent and $-a$ is the slope of the liquidus line on the equilibrium diagram.

The balance between the rate at which solute is rejected at the boundary and the diffusion flux yields

$$\frac{\partial f}{\partial t} = \frac{-aD}{(u_0 - u)(1 - k)\eta}\left\{\frac{\partial c}{\partial y}\right\}_{\eta f} \tag{4.9}$$

where k is the partition coefficient, and

$$f = 0, \qquad t = 0\,. \tag{4.10}$$

---

* See Tayler (p. 135).

The cell size is given by an empirical relationship of the form

$$\eta = F\left\{\frac{\partial u}{\partial t}\right\} . \tag{4.11}$$

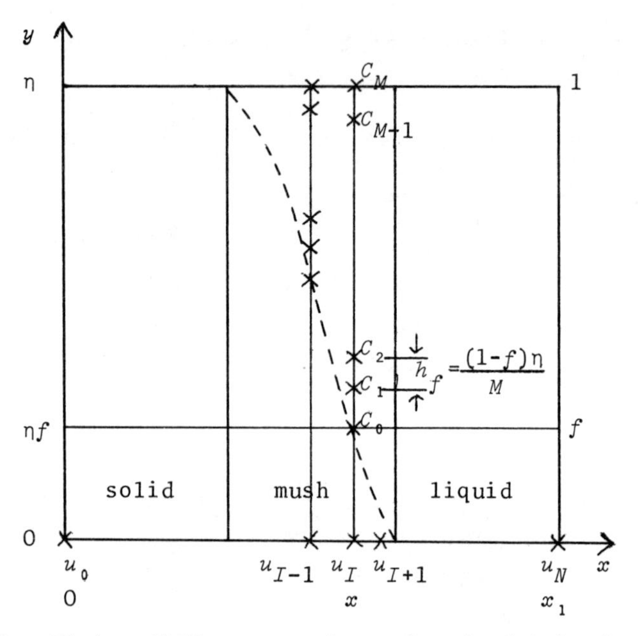

Fig. 7. Finite-difference scheme for dendritic freezing model.

It is easy to see how an explicit method can be used to solve these equations numerically using a finite-difference scheme as in Fig. 7. But several difficulties arise. Firstly, as f tends to unity, so the space step in the y-direction tends to zero and the time step necessary for stability vanishes even more rapidly. An implicit formulation is needed or else an elegant method for reducing the number of space steps from time to time. Again, as f tends to unity so C_0, at the interface, exceeds the eutectic composition C_E, but in practice this cannot happen. It would be possible to cope realistically with this by using b, the slope of the liquidus line on the carbon side of the eutectic in place of $-a$ in (4.8) and (4.9) but it is much simpler and probably just as realistic to restrict all the C's to C_E and shunt any excess down the line.

However, there is one big difficulty which concerns (4.11) defining the cell size as a function of the local solidification rate. Which particular rate should be used? Metallurgists divide evenly between the rate when f first becomes non-zero (*i.e.* at the liquidus) and an average over the whole freezing interval. Peel and Pengelly [4], working on the continuous casting of aluminium, produced evidence however, which implies that it is the final (solidus end) rate which is structure determining. Both this solidus rate and the average rate over the whole freezing range would be difficult if not impossible to incorporate into a transient model like this.

Nevertheless, non-essential changes to (4.1) and its initial and boundary conditions can create models covering several steady state processes of industrial importance such as continuous casting, electro-slag refining and welding. In such models, which incorporate elliptic equations, it is possible to look ahead to determine the rate at the solidus. In all of these processes an ability to obtain a particular type of structure by choosing the operating parameters is becoming increasingly important. The transient model given above, can however be applied to the solid/solid diffusion which accompanies the phase transformations which occur in heat treatment processes. In this case, the cell dimensions are fixed, either by the cast structure alone or as modified by subsequent plastic deformations.

5. OTHER BOUNDARIES

In conclusion, it is worth just mentioning some of the other processes within the industry which have not been given detailed treatment here. About a dozen distinct problems have been identified which were originally posed as having moving boundaries within them, of which three have been described. Like the decarburisation model, one or two more turned out to be pseudo boundary problems. Adequate solutions could be obtained without recourse to moving boundary conditions. Two, more simulations of a coke oven and of a sinter strand, are still at the Class *C* stage and do not as yet yield much of interest in the context of moving boundaries, whilst some others on the engineering side of solidification are described by Perkins (pp. 19-25).

And finally, three other problems although of considerable interest and practical utility, had to be omitted from the discussion since they did not involve heat flow or diffusion. All three are concerned with the changing shapes of the bulk surface of granular materials or of internal interfaces. But this latter topic could fully occupy a whole conference of its own!

REFERENCES

1. Birks, N. and Nicholson, A., "Calculation of Decarburisation Depth in Steels Reheated under Practical Conditions," I.S.I. Publication 123, pp. 219-223 (1970).
2. Carslaw, H.S. and Jaeger, J.C., "Conduction of Heat in Solids," 2nd edition, Oxford University Press (1959).
3. Crank, J., "The Mathematics of Diffusion," Oxford University Press (1956).
4. Peel, D.A. and Pengelly, A.E., "Heat Transfer, Solidification and Metallurgical Structure in the Continuous Casting of Aluminium," I.S.I. Publication 123, pp. 186-196 (1970).

SCRAP MELTING

A. Perkins
(British Steel Corporation)

1. INTRODUCTION

Currently, most Basic Oxygen Furnaces are operated with a charge of 25% cold scrap and 75% hot metal in round terms. The rate at which the scrap melts has a dominant influence on the refining process. It therefore presents an important problem which also has a bearing on other areas. For example, pig iron or scrap is sometimes added to the BOF in the latter stages of refining as a coolant or source of iron. Lump pig iron can be added to the Open Hearth furnace for carbon make-up. The addition of solid ferroalloys is often made in the ladle after tapping the furnace. When continuous casting is used, temperature control is critical and cooling is often effected by dunking slabs or billet additions. In all these processes, it is advantageous to know the rate of melting of the addition and its effects on the bulk temperature of the steel. This knowledge forms an integral part of many static and dynamic control models of BOFs.

2. OUTLINE OF PROBLEM

The most important parameters involved are the differences between carbon contents and initial temperatures of solid and liquid phases, the size and shape of the addition and the state of the fluid flow, which determines the heat transfer.

The process of scrap melting progresses through simultaneous heat and mass transfer.* Heat from the liquid increases the scrap temperature and eventually melts the scrap. Melted iron diffuses into the liquid whilst, most important, carbon diffuses from the liquid into the solid reducing the

*See Hebditch (p.112).

melting point near the solid surface and producing melting.

A simplified version of the scrap melting problem is that of slab dunking. In this case the carbon diffusion aspect is eliminated because we are dealing with a steel slab in liquid steel, both phases having low carbon content. This problem can also be treated as one-dimensional.

Heat transfer from liquid to solid is determined by the heat transfer coefficient. This factor is the most critical unknown in the problem and unfortunately it is the rate determining factor for heat flow. It is difficult to determine either experimentally or by calculation and it is strongly-dependent on the flow characteristics of the liquid phase. Experimental values are typically 3000W/m^{2}K but calculations often give values up to 40,000W/m^{2}K.

3. MATHEMATICAL DETAILS

Several analytical solutions to special case problems exist, see for example Szekely and Themelis [7], but these necessarily involve restrictive assumptions such as constant thermal properties, constant melting point, or sharp phase transition from solid to liquid. The integral profile method of Goodman [4] also suffers from some of these problems and it is difficult to apply if there is non-steady flow in both phases. Further difficulties are introduced when mass transfer is considered.

It is thus tempting to consider a numerical approach using, for simplicity, an explicit finite difference method. Consider first slab dunking where only heat transfer is involved, and where we assume there is no sudden change of heat content at the liquid-solid interface. The classical finite difference form of the heat conduction equation is:

$$\rho c(T_i^{j+1} - T_i^j)/\Delta t = k(T_{i+1}^j + T_{i-1}^j - 2T_i^j)/(\Delta x)^2 \qquad (3.1)$$

where the subscript i refers to the node on the space mesh at a spacing of Δx and the subscript j refers to the time $j\Delta t$.

If we assume that the solid liquid interface is at node n where node n-1 lies in the solid region and we introduce a fictitious node, n+1, then the boundary condition can be written in terms of the heat transfer coefficient h and the temperature of the liquid T_L as:

$$h(T_L^j - T_n^j) = k(T_{n+1}^j - T_{n-1}^j)/(2\Delta x) . \tag{3.2}$$

The fictitious node can now be eliminated by using (3.1) with $i = n$ to give:

$$T_n^{j+1} = T_n^j + \frac{2\Delta t}{\rho c}\left[\frac{h}{\Delta x}(T_L^j - T_n^j) + \frac{k}{(\Delta x)^2}(T_{n-1}^j - T_n^j)\right]. \tag{3.3}$$

We see that beyond the usual stability condition that $\Delta t < 0.5\rho c(\Delta x)^2/k$ we now require $\Delta t < 0.5\rho c(\Delta x)^2/(k + h\Delta x)$.

Provided the mesh is reasonably fine we can take account of the moving boundary by shifting the boundary condition to the node n-1 when

$$0.5(T_n + T_{n-1}) > LIQ \tag{3.4}$$

where LIQ is the temperature at which the solid is assumed to become sufficiently liquid so as to be mixed with the rest of the liquid mass. The method allows the material properties to be functions of temperature.

It is more difficult to find a satisfactory method for determining the temperature of the liquid at the interface. The natural method to choose is to assume that the molten metal is completely mixed and the bulk liquid temperature is determined by a heat balance across the interface. The disadvantage of this method is that it precludes the possibility of the freezing on of a layer of liquid onto the solid in the initial stages of the process, unless the conditions are such that eventually all the bulk metal will freeze.

In order to permit freezing on, it is necessary to consider the temperature in the liquid in more detail than is possible by relying on the heat transfer boundary condition. The simplest method appears to be to insert a boundary layer of liquid metal of thickness $\Delta \ell$ and conductivity k_{ℓ} between the bulk liquid and the solid, where k_{ℓ} is somewhat greater than the conductivity of the liquid metal without convection and is determined either empirically or from a consideration of fluid dynamics. Since the temperature at the boundary layer/bulk metal interface is that of the bulk metal it may still be necessary to difference the boundary layer in order to obtain reasonable estimates of the rate of freezing independent of the time step Δt.

Our work so far has concentrated on the heat transfer with complete mixing model and calculations have been carried out on a range of slab sizes of $3m^2$ cross-sectional area and of thicknesses 0.08m, 0.16m, 0.2m and 0.25m in a ladle of 180 tonnes initially held at 1600°C. Typically a mesh size of 0.01m or less was used, giving a good heat balance, but variations in melting time arose from changes in mesh size and corresponding time step with the coarser meshes. The variations in melting time appeared attributable to the failure to track the moving boundary accurately enough using (3.4) rather than errors in the heat conduction equation. This is understandable in that errors in boundary position are magnified as melting progresses. This suggests that either quite a fine mesh is required or some method of interpolation which positions the boundary continuously between nodes should be used.

The effect of varying the heat transfer coefficient was also investigated for the steel slab in liquid steel when carbon diffusion is unimportant. For a slab of 0.16m thickness the melting time ranged from 53 minutes with a heat transfer coefficient of $2000W/m^2k$ to 18 minutes with a heat transfer coefficient of $7000W/m^2k$, thus confirming that correct modelling of the heat transfer at the interface is critical.

4. CARBON DIFFUSION

Usually in scrap melting the carbon content of the liquid phase is much greater than that of the solid and the diffusion which then takes place produces a significant drop in the melting point of the solid phase. Experimental evidence and practical experience indicate that this is the dominant effect in the determination of melting and cooling rates. Some simplified heat transfer models may allow the incorporation of this factor.

Several attempts, *e.g.* Guthrie and Gourtsoyannis [5] rely on the skin melting approximation. This is applicable for the case in which the flux from the solid surface into the solid interior is negligible compared to the flux into the solid surface from the liquid. The model uses calculated values of heat transfer coefficient appropriate to the agitation in the fluid caused only by natural convection but they are thought to be large enough to permit skin melting. It is not easy to judge the reliability of the skin melting approximation. If it were true then the melting velocity should remain constant. However this is found not to be the case, the velocity increasing as melting proceeds. This is due to two effects: firstly the heating up of the interior (itself a measure of the reliability of the approximation) and secondly, the geometry of the solid which was taken to be spherical. If the two effects were separable we could check the skin melting approximation. A short "skin-freezing" period is predicted ($\sim$ 3 sec), existing until the flux from the fluid exceeds that conducted into the solid. No details are given of how the latter flux is calculated. It would be interesting to know how this difficulty (discussed previously here) has been handled.

The skin melting idea lends itself conveniently to the introduction of carbon diffusion, ([7], p.467). This is a reasonable approach if the difficulties associated with the decrease in melting point and separation of the solidus and liquidus can be overcome. Later developments by Szekely *et al* [8] are more complicated and involve the use of Green's functions.

Work published in the USSR shows clearly the importance of carbon diffusion. Plant trials [1] have shown that scrap added to a 100t converter melts within 12 mins. by which time the liquid metal temperature has risen to only 1480°C, well below the scrap melting point. Theories of diffusion have been proposed subject to simplifying assumptions such as a constant melting velocity [6] or a neglect of the heat transfer coefficient [3].

Two conclusions recur. Carbon diffusion is as important as heat transfer as far as scrap melting is concerned: experimental determinations of heat transfer coefficient give values around 3000W/m^2k.

Several other solutions for scrap melting and related problems have been published, many involving different methods of solution such as implicit finite difference schemes [2].

Finally, it should be added that it has not been the intention of this paper to survey published work or to report very much new work. This paper is presented as bait in the hope of attracting a sufficiently accurate and practical solution of scrap melting and related problems.

REFERENCES

1. Burdakov, D.D. and Varshavskii, A.P., *Stal* (in English) **8**, 647-648 (1968).
2. Crank, J. and Phahle, R.D., *Bull. Inst. Math. and Applic.* **9**, No. 1, 12-14 (1973).
3. Glinkov, M.A., Filmonov, Yu.P. and Yurevich, V.V., "Steel in the USSR," **1**, No. 3, 202-203 (1971).
4. Goodman, T.R., *J.Heat Trans.*, **80**, 335-342 (1958).
5. Guthrie, R.I.L. and Gourtsoyannis, L., *Can. Met. Quart.* **10**, No. 1, 37-46 (1971).
6. Lisin, F.N. and Nevskii, A.S., *Telotizika Vysokikh Temperatur*, **9**, No. 4, 790-795 (1971).

7. Szekely, J. and Themelis, N.J., "Rate Phenomena in Process Metallurgy," Wiley (1971).

8. Szekely, J., Chuang, Y.K. and Hlinka, J.W., *Met. Trans.* 3, No. 11, 2825-2843 (1972).

Meltout and Solidification in Furnace Valves

W.R. Hodgkins and J.F. Waddington
(Electricity Council Research Centre, Capenhurst)

1. INTRODUCTION

In order to improve the versatility of electric metal melting furnaces, a program of work has been undertaken at ECRC to replace the conventional methods of metal removal using a ladle or by tilting, by the use of a nozzle and valve situated either at the bottom or at one side of the furnace. Two designs will be described here. The first is a manually operated nozzle and stopper for a small, up to 4 tonnes, bottom pouring furnace, shown schematically in Fig. 1. The method of operation is that during the heating cycle the metal is heated and stirred by the induced current in the melt, the nozzle being blocked by an air cooled bottom stopper which physically prevents the outflow of metal and also provides sufficient cooling to maintain a solid plug of metal in the nozzle outlet. Before pouring, a top stopper is temporarily inserted, the bottom stopper then removed, and finally the top stopper lifted from its seat to allow egress of the melt. This does not occur instantaneously as time is required to melt the plug of solid metal in the nozzle. One criterion of a good design is that this time should not be too short to allow safe removal of the bottom stopper, nor be too long so that an undue wait is required before pouring begins. To terminate pouring the top stopper is lowered, the bottom stopper is reapplied and finally the top stopper is removed from the furnace.

The second design is a sliding gate valve applicable to larger side-tapped furnaces. A cross-section of the valve and nozzle is shown in Fig. 2. The approach has been to design a valve to fulfil its basic mechanical function and then to submit the proposed design to thermal analysis. The criteria of a satisfactory thermal design are that the time for freezing to begin should be long compared with the expected interval between successive tappings and that the steelwork should be kept below 600°C.

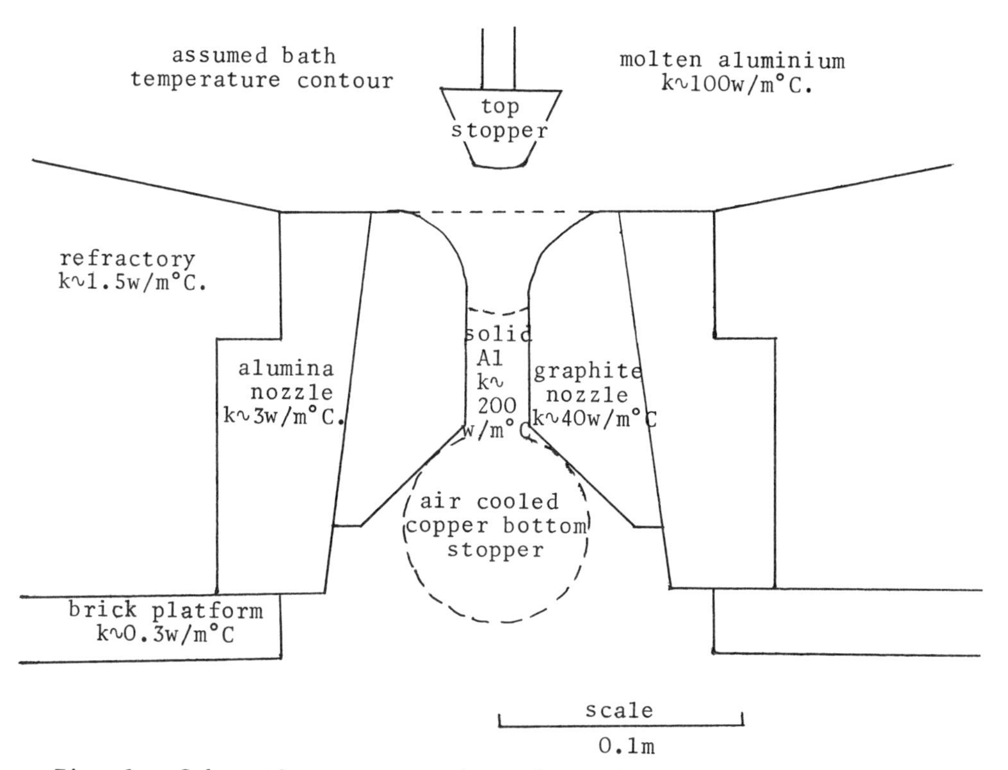

Fig. 1. Schematic cross-section of nozzle region of bottom pouring furnace.

The analyses of both types of valve pose similar problems. Besides allowing for the latent heat over the region of solidification in a transient heat flow problem, the analysis must be able to include a complicated geometry involving several materials with widely different thermal properties. In practice the conductivity and heat transfer data are often not too well determined and therefore over-refinement of the model and mathematical treatment would be superfluous. The program written to tackle these problems has therefore put the emphasis on generality and practicality rather than on unattainable accuracy.

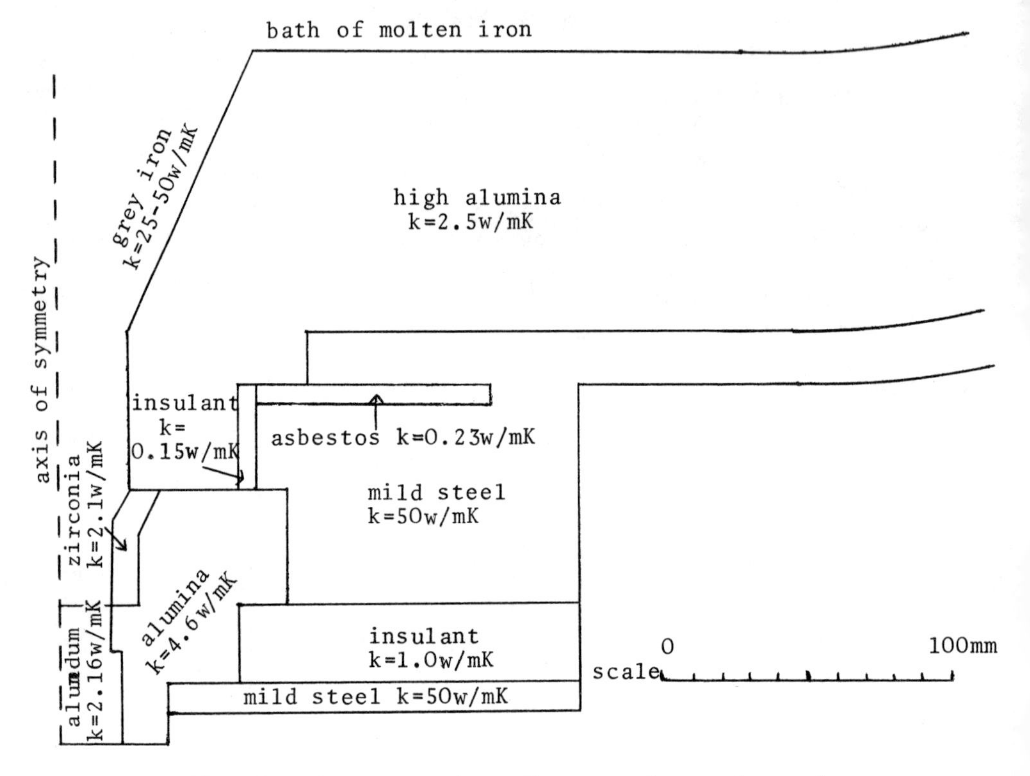

Fig. 2. Cross-section of gate valve and nozzle.

2. PHYSICAL MODEL

To reduce the complexity of the problem it is assumed that the region around the nozzle is axisymmetric. Looked at from the point of view of the metal in the nozzle this should be a reasonable approximation. It is assumed that the thermal conductivity of each material is a known function of temperature. For particular materials, such as extruded graphite it may be anisotropic in which case it is assumed to be defined by at most two components k_r, k_z. For most materials the thermal conductivity is reasonably well known as a function of temperature. However, account must also be taken of the penetration by the molten metal of the surface of the refractory which greatly increases the thermal conductivity of the latter. Rough experimental data are available to define the region of penetration and estimate the thermal conductivity.

In order to include the latent heat of solidification it is convenient to assume that the temperature T is uniquely given as a function of the heat content H. In practice the relation may also depend on such factors as rate of change of temperature, stress, impurities, temperature gradients and past history. The data assumed in the model allow the phase change to occur over a temperature range of several degrees although the numerical method would permit the change to occur over a zero temperature range.

Boundary conditions are imposed by assuming that the bulk molten metal temperature is known and that heat transfer from the furnace to air is known. The latter depends on emissivity and radiative interchange with neighbouring surfaces and cannot be fixed very accurately. The remaining uncertainty relates to the interface between the bulk metal and the solidified metal in the region of the nozzle. It is assumed that in effect the bulk metal has infinite conductivity which means that an interfacial region must be imposed between the solid metal and bulk metal. This is done by drawing an arbitrary boundary across the inside end of the nozzle and ascribing to the molten metal between the boundary and the solidified region a conductivity equal to that of static molten metal. The effect of this assumption will be discussed later.

3. MATHEMATICAL FORMULATION

The basic equation of temperature conductivity for a small volume v surrounded by a surface S states:

$$\frac{\partial}{\partial t}\int_v H\,dv = \int_S k\nabla T.d\underline{S} \tag{3.1}$$

where H is the heat content in joules/m^3, k is the conductivity in watts/m°C and T is the temperature in °C. The data for T as a function H, and k as a function T are contained in tables and linearly interpolated by the program. The present transient program takes k to be a scalar although in general it will be a tensor.

To set up a numerical method of solving the equation it is necessary to choose a discretisation scheme over the space coordinates (r,z) and an integration scheme over the time, t. Because of the irregular geometry and number of different materials a conventional finite difference scheme is inappropriate, whilst a full finite element method would involve time derivatives of H at more than one node on the left-hand side (3.1) and make numerical integration more difficult. A mixed method was therefore chosen in which the region is divided up into a triangular mesh in the (r,z) plane such that material boundaries lie along sides of triangles. The mesh was restricted by the condition that six triangles meet at each interior node, thus forming a topologically uniform triangular mesh. Whilst this has the disadvantage that a maximum of six materials may meet at any node, it has the advantages that it leads to a simpler computer program and could allow the use of "hopscotch" type techniques for the numerical integration. It may also permit the use of isothermal/flux coordinates.

It is assumed that the temperature is linear over each triangle and (3.1) is evaluated over the region around each node formed by the envelope of the perpendicular bisectors of the lines emanating from the node. Whilst this is not the only technique of mesh sub-division, it has the property that it reduces to the conventional finite difference formulae for rectangular meshes; on the other hand some care is required in using obtuse angle triangles as they may give rise to inadmissibly small or even negative volumes at the node. The right-hand side of (3.1) is readily evaluated by summation over the straight line segments surrounding the node, the conductivity being evaluated at the temperature of the mid-point of each triangle side. For an external boundary the heat transfer coefficient is evaluated at the temperature of the point $\frac{1}{4}$ of the distance along the side from the node.

The left-hand side of (3.1) is written in the form:

$$\frac{\partial}{\partial t} \int H \, dv \simeq \Sigma \, \delta v_i \, dH_i \qquad (3.2)$$

where δv_i is the volume of the ith sub-triangle contributing to the node and H_i is the heat content. If all triangles meeting at a node are of the same material then the increment of H is uniquely defined. Otherwise the additional constraint must be added that the increments of temperature in each material at the node are equal. The simplest way to carry this out in practice is to define a composite material at the node and work out initially a $T(H)$ curve for the node. Both sides of the equation are suitably weighted to include the r dependence due to axial symmetry.

The method of numerical integration currently in use is Euler's explicit. There is little point in using a more elaborate method since short time steps are dictated by considerations of numerical stability arising from the presence of some materials with a comparatively high conductivity and the wish to avoid large time steps in the melting region.

Although the same equations can be used to calculate the steady state temperature either by passing through a quasi-transient or by applying successive over-relaxation to the right-hand side of the discretisation of (3.1), this has been found to be very slow and a direct solution of the steady state equation is preferred.

4. RESULTS

Calculations were performed for the bottom pouring furnace containing molten aluminium to determine the length of time the solid plug takes to melt after removal of the bottom stopper. A range of bath temperatures and two different temperatures of the air cooled bottom stopper were investigated. The assumption was made that the bath temperature held on the contour along the top of the nozzle and that below this level there was no mixing in the metal.

The results are presented in Fig. 3 with some experimental results for comparison. It will be seen that the agreement is not very good. Two factors which would act to shorten the calculated times are mixing of the melt in the nozzle and the fact that pouring will start as soon as the solid plug thins to such an extent that it can no longer withstand the pressure head. Alternatively when there is little mixing in the bath, the bath temperature may not be representative of that at the top of the nozzle.

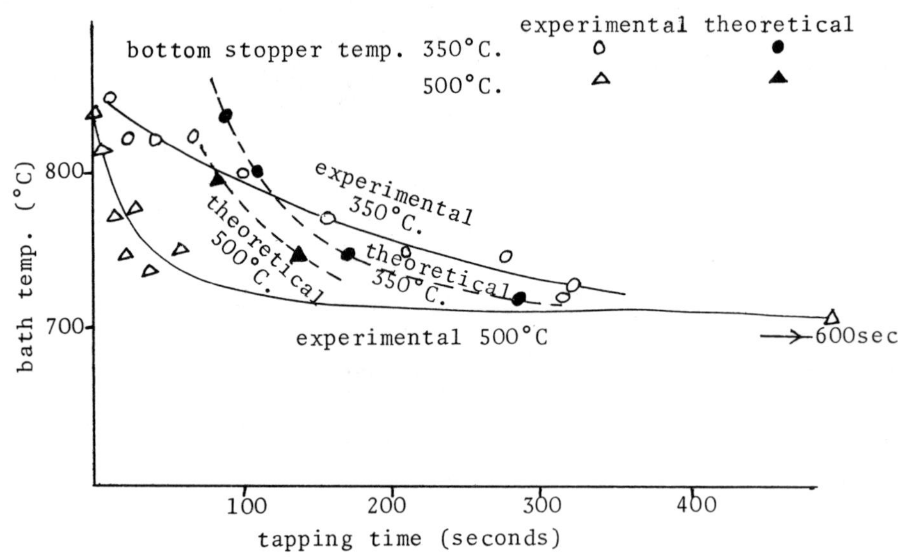

Fig. 3. Graph of bath temperature v. tapping time for bottom pouring furnace.

Because of the discrepancy between theory and experiment some further experimental results were obtained. These too showed poor agreement with the theory and furthermore were inconsistent with the initial experimental results. Investigation of the experimental conditions of furnace operation showed that the operation cycle of the furnace varied depending on individual operator control and that consequently the power, which not only heats the metal but also causes vigorous stirring, was not held at the same level prior to pouring.

Since the purpose of the theory was to compare different nozzle designs for a range of furnace sizes, it was concluded that whilst the theory could be used to provide an order of magnitude guide to the different designs it was inadequate to predict actual melting times. It was clear that in order to improve the model much more information would be required about the degree of mixing in the nozzle region.

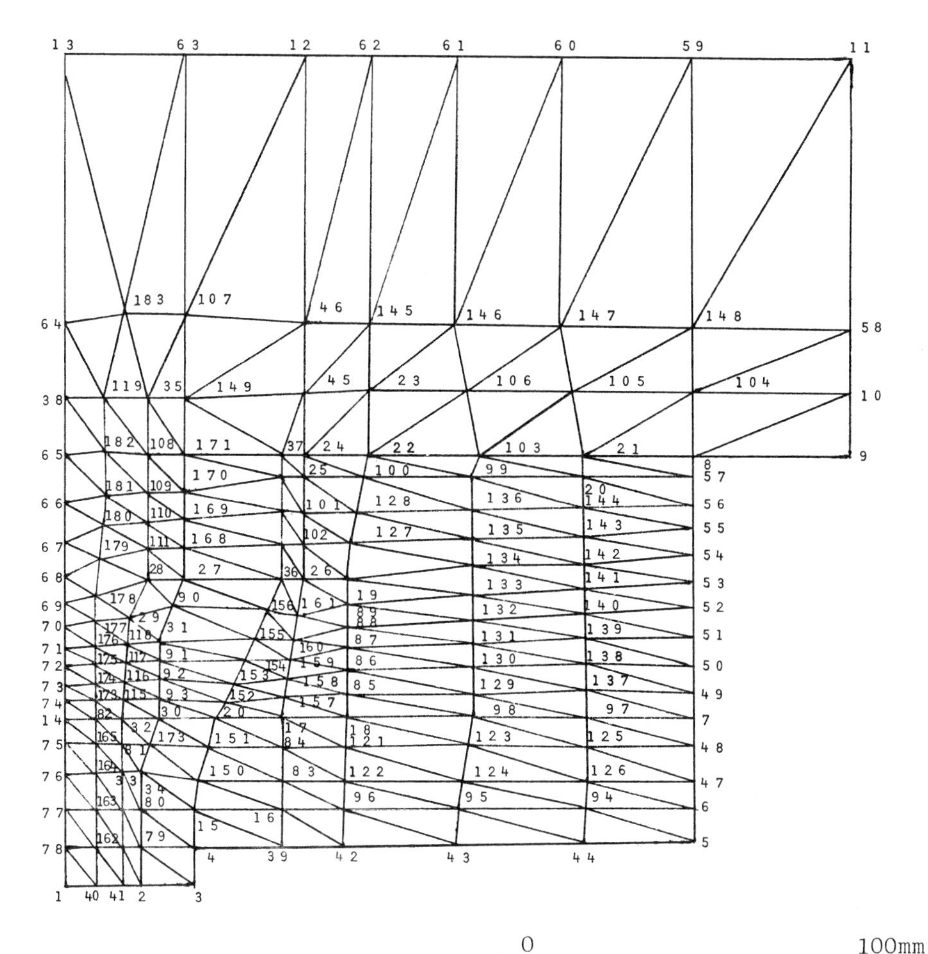

Fig. 4. The triangular mesh used to discretise the heat conduction equation.

A series of calculations were also performed on various modified designs of the sliding gate valve for the side pouring furnace.

The region around the nozzle was discretised as in Fig. 4, and the steady state temperature was first calculated by direct solution for the condition when molten iron is pouring from the furnace at a temperature of 1500°C. A ceramic plug was then inserted in the bottom region of the nozzle at an assumed temperature of 20°C and the transient temperature calculated in the metal for a short time after the insertion. The results for the nodes in the centre of the nozzle are shown in Fig. 5 as a graph of temperature against time. In this case solidification was assumed to occur gradually over the temperature range 1150°C to 1450°C, [2]. In the design quoted here solidification starts after about $4\frac{1}{2}$ seconds and the design was later modified to prevent solidification. Once again the problem arises of the reasonableness of the assumption that the bath temperature prevails at the inside edge of the nozzle and that there is no mixing within the nozzle region. However, in this case the design is likely to be conservative as mixing within the nozzle will help to prevent freezing inside the valve.

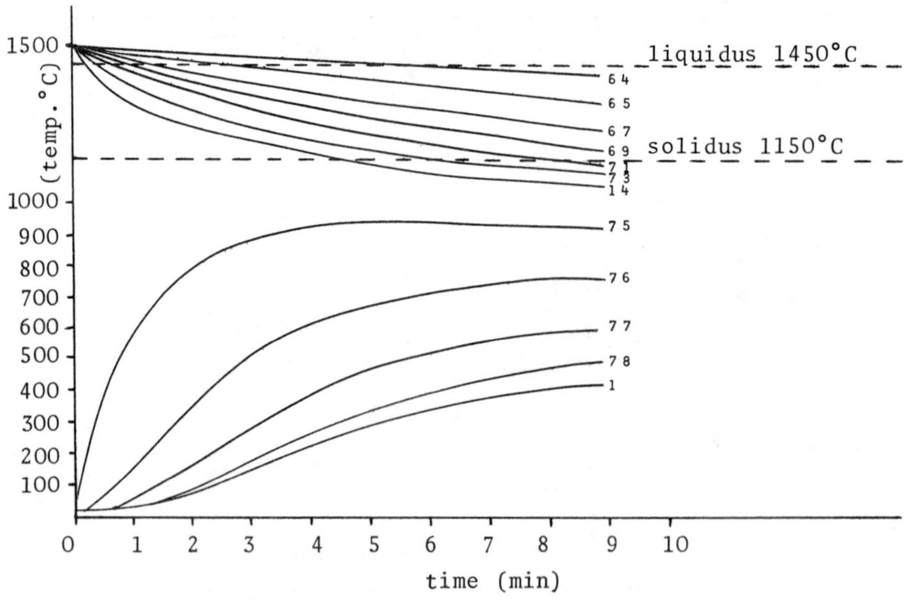

Fig. 5. Graphs of time against temperature for axial nodes in the case of melting over a temperature range.

Whilst the assumption that solidification occurs over a temperature range rather than at a fixed temperature is true of the actual melt (grey iron), it is interesting to see the effect on the numerical results of assuming a definite solidification temperature of 1200°C. The results are shown in Fig. 6 and it is evident that there are a series of steps in the temperature against time curves which appear to correspond to the fact that each node controls a finite amount of material, thus forcing the left-hand side of (3.1) to behave in a stepwise manner. If smoother curves are desired it is possible to either smooth the resultant curves directly, to reduce the mesh size, or to smear out the solidification region over a finite temperature range. The latter has been recommended by Rubinstein [3].

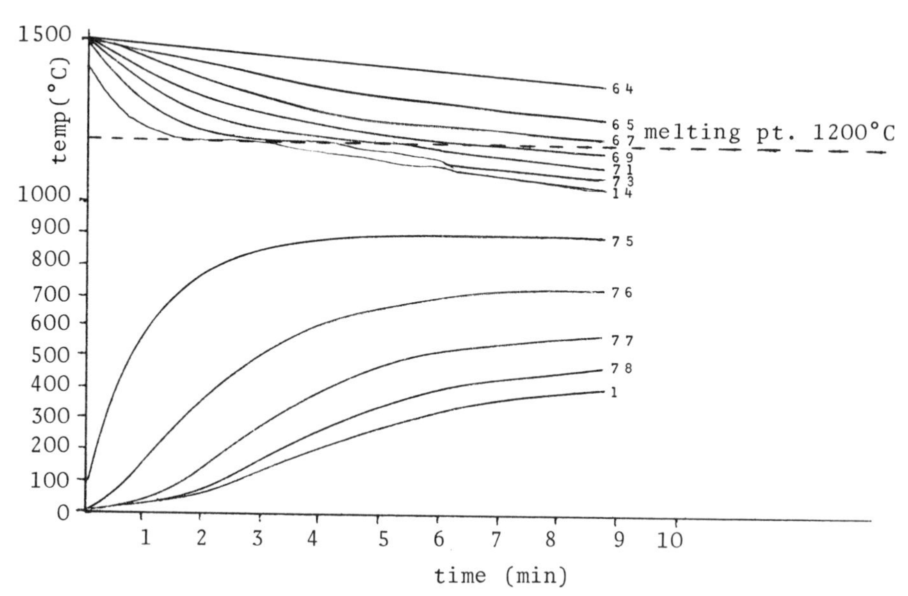

Fig. 6. Graph of time against temperature of axial nodes for the case of a sharp melting point.

Although in principle higher accuracy and smoothness could be obtained by reducing the mesh size, this option is only of limited value as for the particular mesh size used the time step was restricted by stability of the Euler explicit method to less than one second and any further reduction in mesh size without the introduction of some other compensating feature would lead to inordinately long and expensive integration times. The reason for the very short time step is that it is necessary to divide a material with a high thermal conductivity (solid iron) into a fine mesh in order to determine the surface of solidification. This yields a very quick response time for the nodes, and yet the total time of interest is influenced by the slow response of the total system resulting from its large thermal mass and the presence of insulating material.

5. DISCUSSION

The results show that it is possible to model the problem of the furnace valve using a basically simple numerical technique. As far as obtaining correct results the major problem lies in modelling correctly the boundary conditions at the liquid-solid interface or in otherwise specifying the thermal transport of the liquid region. Mathematically this is a completely different problem from the one considered, but it nevertheless is believed to contribute more uncertainty to the results than any other factor.

The numerical method chosen appears to be moderately successful at dealing with the phase change and the results appear to be significant. In situations where there is a definite melting point it is still possible to define the interface to an accuracy of better than one mesh length either by smoothing the temperature curves and interpolating or by allowing the phase change to occur over a temperature range. Even with a numerical method which follows the phase interface, there is still a numerical error in position or temperature and hence there is little point in defining the phase change temperature range much more accurately than that of the eventual solution.

The most severe restriction of the present numerical method is not specifically concerned with the phase change. It is the small time step required to preserve stability using the Euler explicit integration technique. This aspect warrants further investigation. Use of implicit techniques often requires so much computation at each time step as to negate the advantages of numerical stability for large time steps and it may well be that stable explicit schemes, such as versions of the "hopscotch" technique, may provide the answer. The convergence of implicit methods for multidimensional Stefan problems has been considered in some detail by Meyer [1].

REFERENCES

1. Meyer, G., *SIAM J. Numer. Anal.* **10**, No. 3, 522-538 (1973).

2. Ross, R., "Metallic Materials Specification Handbook," Spon. (1972).

3. Rubinstein, L., "The Stefan Problem," (Translation of Mathematical Monographs **27**), *Amer. Math. Soc.* (1971).

ANALYTICAL AND NUMERICAL TECHNIQUES FOR ABLATION PROBLEMS

J.G. Andrews and D.R. Atthey
(Central Electricity Generating Board, Marchwood Engineering Laboratories)

1. INTRODUCTION

The cutting and welding of metals is of major importance in many areas of technology. Whilst the saw and the welding torch are still the commonest tools for these jobs, there is considerable interest in devising new techniques especially for special materials or where some degree of automation is desirable. In recent years there have been attempts to develop high power lasers and electron beams for both cutting and welding [3], [9], [14] and [16].

The essential idea is to focus a lot of power on to a small area of the surface of a metal, thereby producing intense surface heating and evaporation and the subsequent formation of a hole (see Fig. 1). In cutting one tries to arrange conditions such that a hole is drilled right through the material without molten metal slopping back into the hole and resolidifying. In welding the situation is reversed - one butts two pieces of metal together, heats them along their join so that the molten metal from both sides intermixes and then resolidifies when the heat source is taken away.

In this paper we consider a mathematical model of central interest in deep penetration welding and cutting which attempts to answer the question of how fast is it possible to dig a hole with a high power beam. For the case of a beam of constant power it is shown that the motion of the evaporating boundary can be obtained using a perturbation technique based on the fact that the latent heat is large compared with the heat capacity of the material. The case of a beam with a time-varying power density is treated numerically using a modification to Douglas and Gallie's [6] implicit finite difference method.

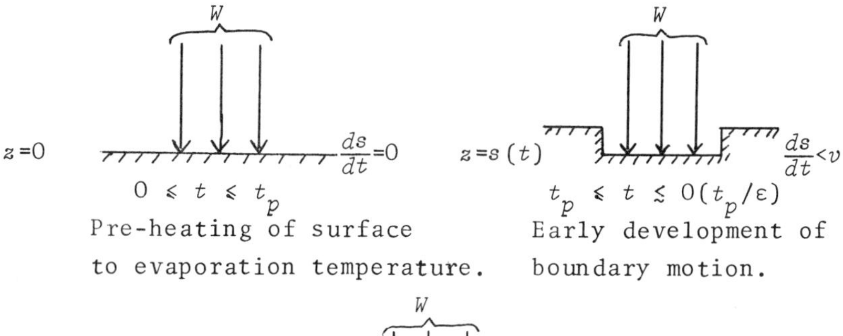

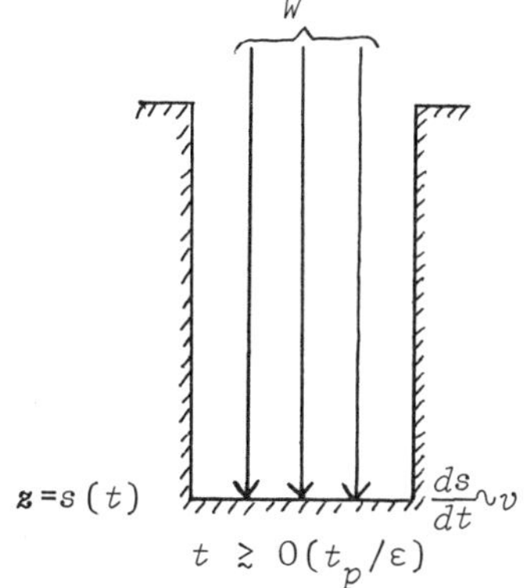

Fig. 1. Various stages of hole formation.

2. BASIC PHYSICAL MODEL FOR ABLATION

Consider a high power laser beam or electron beam striking a small area of a metal surface. Some of the energy is absorbed and the rest is reflected. The absorption of energy takes place within a thickness typically much less than a millimetre, so that there is surface heating and the surface temperature rises. It does not rise indefinitely. There are two processes which limit it. The first is heat conduction into the material from hotter to colder metal. The second is evaporation. When a material boils, *latent* heat is absorbed without any further rise in temperature as the material vaporises. As the vapour puffs away from the surface, so a hole develops in the material and it is the need for a quantitative description of this process that calls for mathematical modelling.

3. EVAPORATION CONTROLLED LIMIT

The simplest case to consider is that in which all the energy applied at the surface is used to evaporate the material. This "*evaporation-controlled limit*" can arise in two ways. The first is that where the energy is applied to the surface too rapidly for heat to be conducted into the material. The second is that where the beam power density is constant and the temperature distribution ahead of the evaporating boundary approaches a steady state after a sufficiently long time.

Assume that the power, W, is distributed uniformly over some area of the surface, A, and is applied normally to the surface. In a time interval δt the amount of energy dissipated is $W\delta t$ and the depth of the hole increases by δs, say. Hence the volume of material evaporated is $A\delta s$ and by conservation of energy, we have

$$h\rho A\delta s = W\delta t$$

where h is the heat required to vaporise unit mass of material and ρ is the density of the material. Rearranging and letting $\delta t \to 0$, we obtain the speed at which the hole develops as

$$ds/dt = (W/A)/h\rho \quad . \qquad (3.1)$$

Equation (3.1) shows that, for any given material, the limiting speed is proportional to power density, W/A. For example, for a beam power density of 1 kW/mm^2 on steel, the limiting speed is 17 mm/s. Integrating (3.1) and setting $s = 0$ at $t = 0$, the depth of the hole at any instant t is

$$s(t) = (1/h\rho A)\int_0^t W dt \quad ,$$

or

$$s(t) = E(t)/h\rho A \quad ,$$

where $E(t)$ is the total energy dissipated by the source in the time interval $(0,t)$. Thus, in the evaporation-controlled limit, the depth of the hole depends only on the total energy supplied to the surface.

4. ALLOWANCE FOR HEAT CONDUCTION

In practice there will always be some conduction of heat into the material, so the evaporation-controlled speed (3.1) represents an upper limit on the rate of penetration and it is of interest to calculate the characteristic time to approach this limit. The general problem of the motion of a phase boundary with heat conduction is known as the Stefan problem and is well known to be difficult mathematically [4] and [13].

Let us suppose initially that heat is conducted normal to the surface. Essentially we have to solve the one-dimensional unsteady heat conduction equation [4]

$$\partial^2 T/\partial z^2 = (1/D)\,\partial T/\partial t \tag{4.1}$$

for the temperature $T(z,t)$ inside the material where $D = K/\rho c$ is the thermal diffusivity, K, ρ and c are the thermal conductivity, density and specific heat, respectively, subject to boundary conditions at the moving boundary $z = s(t)$ and at the far face of the material.

One boundary condition at the moving boundary is obtained by applying energy conservation there, *i.e.*

(rate of energy absorption by surface)	=	(rate of energy conversion into latent heat of evaporation)	+	(rate of heat conduction into the material)

i.e.

$$F = L_v \rho (ds/dt) - K\partial T/\partial z \tag{4.2}$$

where $F = W/A$ is the power density and L_v is the latent heat of evaporation per unit mass. Another boundary condition is that the temperature of the moving boundary is approximately equal to the boiling point, so that

$$T = T_v \tag{4.3}$$

on $z = s(t)$. In practice, for all materials, there is always a certain amount of evaporation below the actual boiling point but the vapour pressure is insignificant compared with atmospheric pressure except very close to the boiling point, so that (4.3) is a fair approximation.

For penetrating thick materials the presence of the far face is relatively unimportant and it is reasonable (and convenient) to remove the far face to infinity, where we put

$$T = 0;$$

(we take the ambient temperature of the material to be zero without loss of generality). For completeness it is also necessary to state boundary conditions at the other phase boundary *i.e.* between the solid and the liquid. However, for many materials of interest the ratio of the latent heats of fusion and evaporation is small compared with unity and the discontinuity at the melting boundary can be ignored to a good approximation.

This Stefan problem has the complication (compared with the classical Stefan problem) of a power term at the moving boundary and the usual approach to solution would be by means of a numerical method [8], [10], [11] and [12]. However, when the power density is constant, it is possible to develop a perturbation solution. Now the ratio of the heat lost by conduction to that by latent heat of evaporation is

$$|K(\partial T/\partial z)/\{L_v \rho (ds/dt)\}| = K\,O(T_v/l)/L_v \rho\, O(W/Ah\rho) \quad ,$$

using (3.1), where l is some characteristic length for temperature decay into the material in the solution of the steady state heat conduction equation in moving coordinates. A suitable choice of l is

$$l = D/(W/Ah\rho) \quad . \tag{4.4}$$

Substituting for l and D, we have

$$|K(\partial T/\partial z)/L_v \rho (ds/dt)| = O(\varepsilon) \quad ,$$

where

$$\varepsilon = cT_v/L_v \tag{4.5}$$

is a constant for the material which is typically small compared with unity for many materials of interest (listed in Table I). The smallness of ε suggests a perturbation treatment of the case with constant power density with the simple solution in the evaporation-controlled limit forming the zero order approximation.

On the other hand, the case of a time-varying power density will generally be unmanageable analytically and we must resort to a numerical solution (see Section 6).

5. PERTURBATION SOLUTION FOR CONSTANT POWER DENSITY

5.1 Simple Perturbation Solution

We now develop a simple perturbation solution for the constant power density source in terms of our small parameter, $\varepsilon = cT_v/L_v$, to determine the motion of the boundary after long times.

In order to distinguish different orders of approximation it is convenient to introduce the following normalised variables

$$\theta = T/T_v, \quad \zeta = z/l, \quad \xi = s/l \quad . \tag{5.1}$$

Taking v to be the speed of the boundary in the evaporation-controlled limit we have from (3.1), setting $F = W/A$

$$\begin{aligned} v &= F/(cT_v + L_v)\rho \\ &= (F/L_v\rho)/(1 + \varepsilon) \quad . \end{aligned} \tag{5.2}$$

The characteristic length, l, is defined by (4.4). The normalised heat conduction equation (4.1) then becomes

$$\partial^2\theta/\partial\zeta^2 = \partial\theta/\partial\tau \tag{5.3}$$

with

$$\{(d\xi/d\tau) - 1\} - \varepsilon\{(\partial\theta/\partial\zeta) + 1\} = 0 \tag{5.4}$$

and

$$\theta = 1 \tag{5.5}$$

on the moving boundary

$$\zeta = \xi(\tau) \quad . \tag{5.6}$$

Material	Heat Capacity (from 0°C) cT_v (kJ/kgm)	Latent Heat of Fusion L_f (kJ/kgm)	Latent Heat of Evaporation L_v (kJ/kgm)	Ratio of Latent Heats L_f/L_v	ε cT_v/L_v
Aluminium	2490	389	10800	0.036	0.23
Carbon (Graphite)	3350	-	59100	-	0.06
Copper	962	205	4770	0.043	0.20
Gold	368	66.9	1740	0.039	0.21
Iron	1290	272	6070	0.045	0.21
Lead	217	25.0	861	0.029	0.25
Magnesium	1150	364	5610	0.065	0.21
Mercury	41.9	11.7	305	0.038	0.14
Nickel	1260	301	6361	0.047	0.20
Tungsten	690	(256)	4020	(0.063)	0.17
Water	419	335	2260	0.15	0.19
Zinc	340	109	1780	0.045	0.19

Table I.

Let us try the following perturbation series for θ, ξ

$$\theta(\zeta,\tau) = \theta_0(\zeta,\tau) + \varepsilon\theta_1(\zeta,\tau) + \ldots \tag{5.7}$$

$$\xi(\tau) = \xi_0(\tau) + \varepsilon\xi_1(\tau) + \ldots \quad . \tag{5.8}$$

Substituting for θ and ξ in equations (5.3) to (5.5) yields to zero order

$$\partial^2\theta_0/\partial\zeta^2 = \partial\theta_0/\partial\tau, \; d\xi_0/d\tau = 1, \; \theta_0 = 1 \; . \tag{5.9a,b,c}$$

Ignoring pre-heating effects while the boundary is being raised to its boiling point, the zero-order solution for ξ_0 is obtained from integration of (5.9b) as

$$\xi_0 = \tau \tag{5.10}$$

and that for θ_0 can be obtained directly by first transforming to the frame of reference moving with the boundary and then taking the Laplace transform, yielding

$$\theta_0 = \tfrac{1}{2}e^{-(\zeta-\tau)}\mathrm{erfc}\{\tfrac{1}{2}(\zeta - \tau)/\tau^{\frac{1}{2}}\} + \tfrac{1}{2}\mathrm{erfc}\{\tfrac{1}{2}\zeta/\tau^{\frac{1}{2}}\}. \tag{5.11}$$

The first order equations are

$$\partial^2\theta_1/\partial\zeta^2 = \partial\theta_1/\partial\tau, \; d\xi_1/d\tau = 1 + \partial\theta_0/\partial\zeta, \; \theta_1 = 0. \tag{5.12a,b,c}$$

Differentiating θ_0 with respect to ζ in (5.11) and substituting in (5.12b) gives

$$d\xi_1/d\tau = \tfrac{1}{2}\mathrm{erfc}(\tfrac{1}{2}\tau^{\frac{1}{2}}) - (\pi\tau)^{-\frac{1}{2}}e^{-\frac{1}{4}\tau} \; . \tag{5.13}$$

Combining (5.9b) and (5.12b), we obtain the corrected normalised speed of the moving boundary as

$$d\xi/d\tau = 1 + \varepsilon\{\tfrac{1}{2}\mathrm{erfc}(\tfrac{1}{2}\tau^{\frac{1}{2}}) - (\pi\tau)^{-\frac{1}{2}}e^{-\frac{1}{4}\tau}\} + O(\varepsilon^2). \tag{5.14}$$

5.2 The Problem for Early Times

In Section 4 we obtained a solution which is a good approximation for large times, *i.e.* $\tau > 1$. For $\tau \ll 1$ there are two difficulties. One problem is that our simple perturbation solution breaks down for $\tau = O(\varepsilon^2)$ when the boundary is moving much slower than the evaporation-controlled limiting speed given by (3.1). A second problem arises from the fact that the front face of the material will initially be below the boiling point, T_v. Using Laplace transforms, it is not difficult to show that the time to "pre-heat" the surface while it is stationary up to the boiling point is

$$\tau_p = \tfrac{1}{4}\pi\varepsilon^2/(1 + \varepsilon)^2 \qquad (5.15)$$
$$= O(\varepsilon^2)$$

and that the temperature profile at the end of the pre-heating time is [4]

$$\theta_p = \exp[-\{\zeta(1+\varepsilon)/\pi^{\frac{1}{2}}\varepsilon\}^2] - \{\zeta(1+\varepsilon)/\varepsilon\}\operatorname{erfc}\{(1+\varepsilon)/\pi^{\frac{1}{2}}\varepsilon\}. \quad (5.16)$$

The details of the correction to the simple perturbation solution (given above), in order to describe the early motion of the boundary for times of $O(\varepsilon^2)$ are given in Andrews and Atthey [1]. Matching the two solutions together, using van Dyke's matching principle [17], a uniformly valid solution for the speed of the boundary is

$$d\xi/d\tau = [1+\varepsilon\{\tfrac{1}{2}\operatorname{erfc}(\tfrac{1}{2}\tau^{\frac{1}{2}}) - (\pi\tau)^{-\frac{1}{2}}e^{-\frac{1}{4}\tau}\}][(2/\pi)\{1-\varepsilon/(\pi\tau)^{\frac{1}{2}}\}] \times \sin^{-1}\{(1 - \tfrac{1}{4}\pi\varepsilon^2/\tau)^{\frac{1}{2}}\} .$$

It has errors of $O(\varepsilon)$ when $\tau = O(\varepsilon^2)$ and of $O(\varepsilon^2)$ when $\tau = O(1)$.

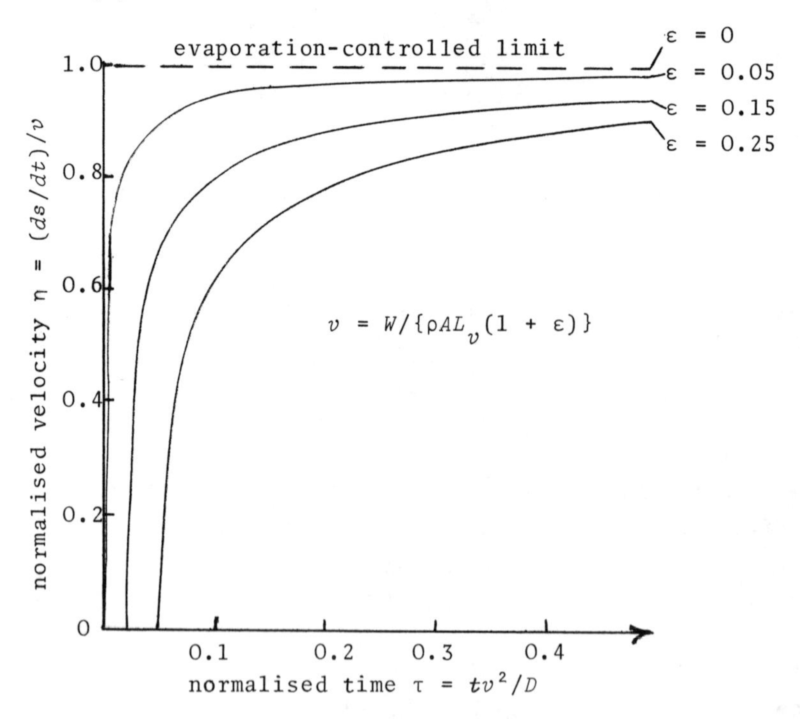

Fig. 2. Variation of (normalised) velocity of evaporating boundary with (normalised) time.

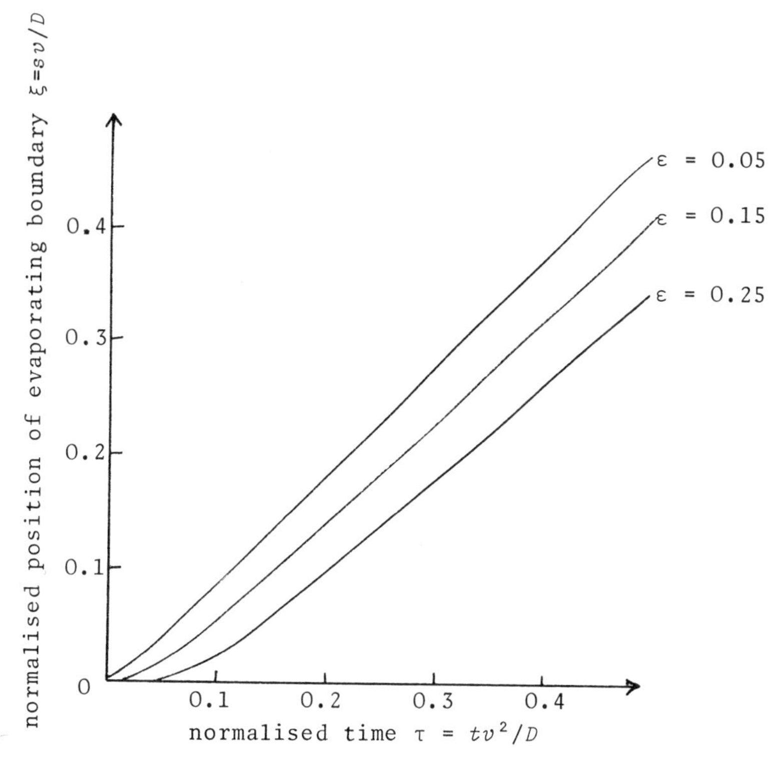

Fig. 3. Variation of (normalised) position of evaporating boundary with (normalised) time.

5.3 Results of Simple Perturbation Solution

Figs. 2 and 3 show how the velocity and the position of the boundary vary with time for typical values of the ratio $\varepsilon = cT_v/L_v$.

The temperature distributions at various times are shown in Fig. 4 for $\varepsilon = 0.2$. (N.B. Pre-heating ends at $\tau = 0.0314$.)

It is perhaps remarkable that a perturbation method can be used for the Stefan problem of evaporation with a power term present since such methods are rarely possible for the classical Stefan problem, in which no power term is present. The explanation lies in the boundary condition (4.2) at the evaporating surface. The boundary condition contains three terms and for the essential part of the problem the conduction term $-K(\partial T/\partial z)$ is much less than either of the other two terms, and can therefore be neglected to a first approximation. In the classical Stefan problem the heat conduction term is one of the only two terms which appear in the boundary condition and both terms must of course be of the same magnitude and it is then not possible to ignore the heat conduction term. However, Friedman [7] has shown that perturbation techniques can indicate the aysmptotic behaviour of the classical Stefan problem for large times.

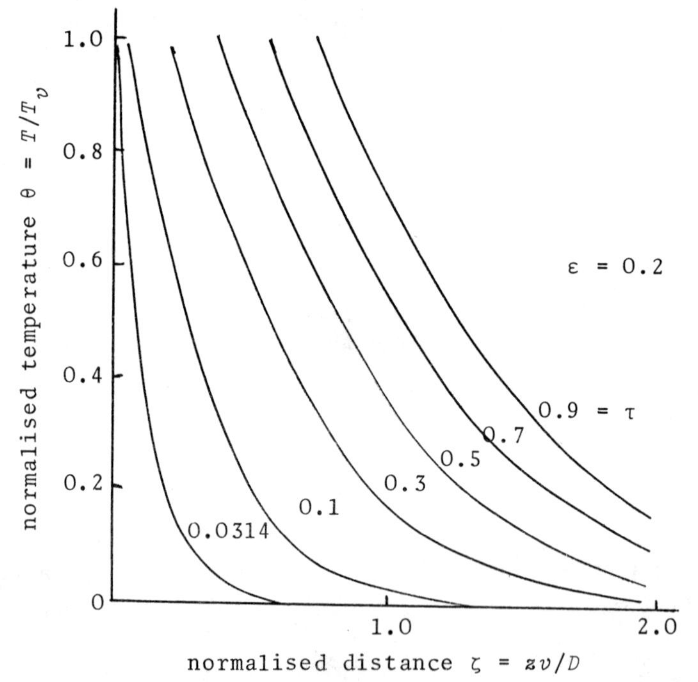

Fig. 4. (Normalised) temperature profiles at various (normalised) times.

The perturbation treatment used in this section can be applied to other problems of practical interest.* Tadjbakhsh and Liniger [15] used a similar technique for a problem in dip soldering. Another problem of technological importance in welding and cutting is that where the source moves with a constant transverse velocity across the material. The depth of the hole is then finite and energy balance considerations will almost certainly yield an upper limit to depths of

*See Ockendon (p.145).

penetration that can be achieved in practice. Other factors may significantly limit the ultimate depth, especially for low transverse velocities, such as attenuation of the beam by the vapour, which may be ionised, and liquid motion inside the hole.

A more complete discussion of the perturbation technique is given in Andrews and Atthey [1].

6. NUMERICAL SOLUTION FOR VARIABLE POWER DENSITY

The perturbation technique described in Section 5 is unwieldy for pulsed sources. For pulses of arbitrary shape it is more convenient to use a numerical method.

We have chosen for the present method to modify Douglas and Gallie's [6] implicit method to include the power term at the phase boundary. Our method employs the Crank-Nicolson implicit finite difference scheme [5], which has a superior convergence rate (of order $\Delta t^2 + \Delta z^2$) compared with finite difference schemes for other parabolic systems. In this method the spatial mesh is uniform and the time step during the evaporation process is chosen such that the phase boundary moves through a complete spatial increment.

The boundary condition (4.2) is not convenient for our numerical scheme when the boundary is moving. A more appropriate form is simply obtained by considering the total rate of change of heat in the material

$$(d/dt)\int_{s(t)}^{\infty} \rho c T(z,t)\,dz = \int_{s(t)}^{\infty} \rho c(\partial T/\partial t)\,dz - \rho c T_v(ds/dt)$$

$$= \int_{s(t)}^{\infty} K(\partial^2 T/\partial z^2)\,dz - \rho c T_v(ds/dt)$$

using equation (4.1) and the relation $D = K/\rho c$,

$$= K\left[\partial T/\partial z\right]_{s(t)}^{\infty} - \rho c T_v(ds/dt)$$

$$= -K(\partial T/\partial z)_{z=s(t)} - \rho c T_v(ds/dt) \quad ,$$

so that

$$(d/dt)\int_{s(t)}^{\infty} \rho c T(z,t)\,dz = F(t) - \rho(cT_v + L_v)(ds/dt) \quad . \quad (6.1)$$

Integrating (6.1) over any time interval (t_1, t_2) we obtain

$$\int_{t_1}^{t_2} F(t)\,dt = \rho(cT_v + L_v)\{s(t_2) - s(t_1)\}$$
$$+ \rho c\int_{s(t_2)}^{\infty} T(z,t_2)\,dz - \rho c\int_{s(t_1)}^{\infty} T(z,t_1)\,dz, \quad (6.2)$$

which can be used as the basis of an iterative loop.

Throughout the finite difference scheme we use a fixed value of Δz. Whilst the material is evaporating, the time step is chosen as the time taken for the phase boundary to move through a distance Δz. Hence in one time step the phase boundary moves from one grid point to the next. On the other hand, when the surface is below the boiling point (and is therefore stationary) a fixed value of Δt is used. In both cases the unsteady heat conduction equation (4.1) is replaced by a Crank-Nicolson finite difference scheme.

When the surface is below the boiling point the boundary condition (4.2) (with $ds/dt = 0$) is replaced by a simple finite difference approximation and the resulting tri-diagonal system of equations is solved using Gauss' elimination method.

When the surface is at the boiling point (and is moving) an initial value of Δt is chosen and the Crank-Nicolson finite difference scheme is solved subject to the boundary condition (4.3). A finite difference approximation to (6.2) then yields a revised estimate for the time step Δt; this iteration is repeated until the desired accuracy is achieved before continuing to the next time step. Fig. 5 compares the analytical and numerical solutions for the case of constant power density $\varepsilon = 0.2$; the difference in the two solutions is around 4% of the final velocity, *i.e.* of $O(\varepsilon^2)$.

Fig. 6 compares the penetrations achieved by pulses with different on/off ratios but the same average power. There are two limiting cases: continuous power and infinitely sharp spikes (which can be described mathematically by delta functions). Maximum penetration is achieved with infinitely sharp pulses since all the power is used for evaporation and none is lost by heat conduction into the material. Penetration decreases continuously as the on/off ratio increases and more heat is conducted into the material. From the analytical solution for the continuous power case we know that the velocity of the phase boundary approaches the evaporation-controlled limiting velocity in a time of order $2K\rho T_v L_v(1 + \varepsilon)^2/(W/A)^2$ (typically 0.5 second for a power density of 1kW/mm^2 with steel), so the reduced penetration for larger on/off ratios is due to the larger amount of heat lost by conduction into the material during this warm-up period. A more detailed description of the numerical method used here is given by Andrews, Atthey and Cross [2].

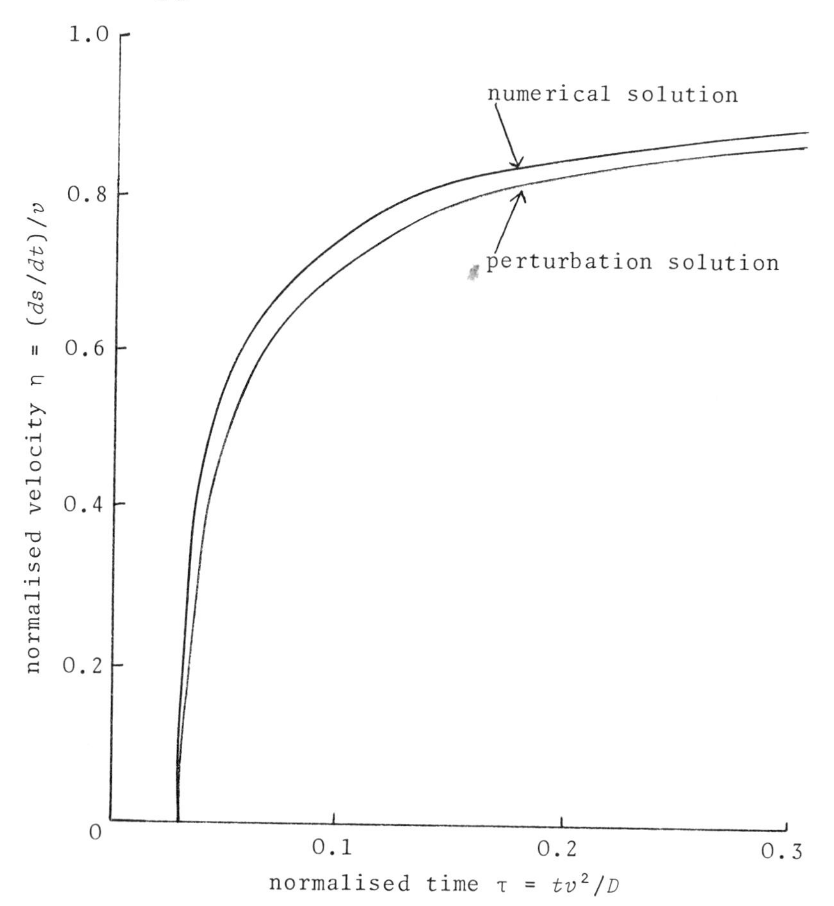

Fig. 5. Variation of (normalised) speed of evaporating boundary with (normalised) time for a constant power density beam.

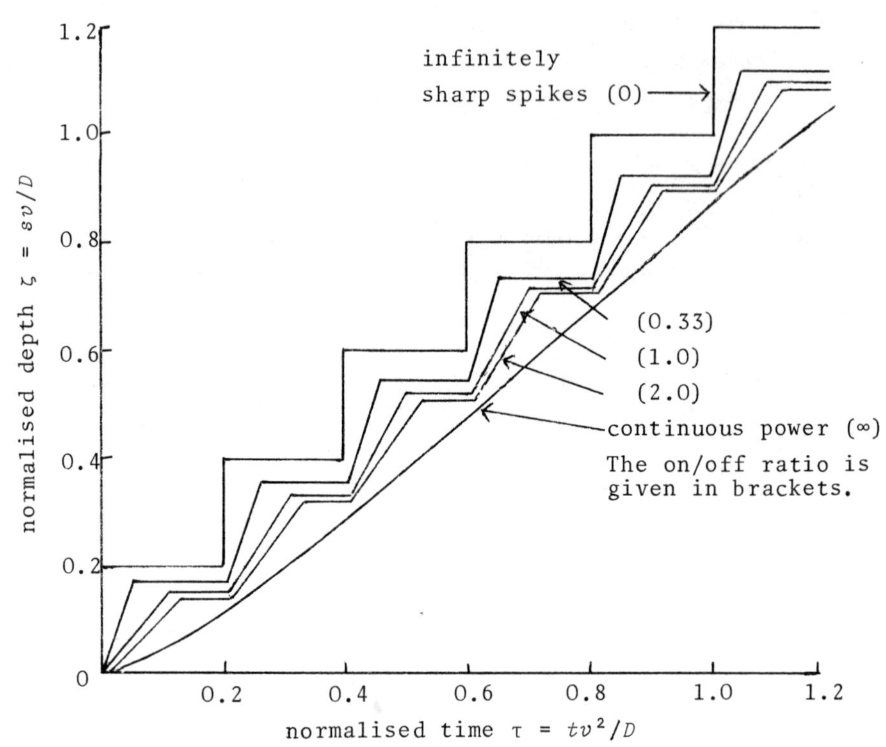

Fig. 6. Variation of (normalised) depth of penetration with (normalised) time for various square wave pulses of same mean power.

REFERENCES

1. Andrews, J.G. and Atthey, D.R., CEGB Laboratory Note R/M/N/689 (also to be published in *JIMA*) (1974).

2. Andrews, J.G., Atthey, D.R. and Cross, A.D., CEGB Laboratory Note R/M/N/723 (1974).

3. Arata, Y. and Miyamoto, I., *Trans. Japan Welding Soc.* **3**, 143-180 (1972).

4. Carslaw, H.S. and Jaeger, J.C., "Conduction of Heat in Solids," 2nd edition, Clarendon Press, Oxford (1959).

5. Crank, J. and Nicolson, P., *Proc. Camb. Phil. Soc.* **43**, 50-67, 235, 238-246, 248, 250-251 (1947).

6. Douglas, J. and Gallie, T.M., *Duke Math. J.* **22**, 557-570 (1955).

7. Friedman, A., "Partial Differential Equations of the Parabolic Type," Prentice Hall, NJ. (1964).

8. Landau, H.G., *Q. Appl. Maths.* **8**, 81-94 (1950).

9. Locke, E.V., Hoag, E.D. and Hella, R.A., *IEEE J. Quant. Elect.* **QE-8**, 132-135 (1972).

10. Lotkin, M., *Q. Appl. Maths.* **18**, 79-85 (1959).

11. Masters, J.I., *J. Appl. Phys.* **27**, 477-484 (1956).

12. Ready, J.F., *J. Appl. Phys.* **36**, 462-468 (1965).

13. Rubinstein, L.I., "The Stefan Problem," *Amer. Math. Soc. Trans. of Mathematical Monographs* **27** (1971).

14. Rykalin, N.N. and Uglov, A.A., *High Temp. Amer. Inst. Phys.* (Consultants Bureau) **9**, 522-527 (1971).

15. Tadjbakhsh, I. and Liniger, W., *Quart. J. Mech. Appl. Math.* **17**, 141-155 (1964).

16. Tong, H., "Heat Transfer and Cavity Penetration During Electron Beam Welding," (University of California) (1969).

17. Van Dyke, M., "Perturbation Methods in Fluid Mechanics," Academic Press, NY. (1964).

A Numerical Method to Determine the Temperature Distribution Around a Moving Weld Pool

D. Longworth

(Central Electricity Generating Board, Marchwood Engineering Laboratories)

1. INTRODUCTION

To further the development of automated welding techniques it has become important to be able to relate the temperature distribution, and hence weld pool shape, to the various welding parameters such as energy input rate and speed of welding. This is to enable weld pools of adequate size to be produced without complete penetration which would lead to hole formation in the welded specimen. Also of interest is the behaviour of the weld pool under transient conditions such as encountered with pulsed energy input since oscillations of the weld pool boundary can produce an alignment of the material grains resulting in a better quality of weld.

In most cases of practical interest geometries are complex and material properties are temperature dependent. Thus a numerical approach is indicated. This paper describes such a numerical method for calculating the temperature distribution around a moving volume energy source. Latent heat effects at the weld pool boundary have an influence on the shape of the pool. The method to be described allows latent heat effects to be introduced instantaneously at the melting temperature, as in the case of a pure material, or gradually over a temperature range as is usually the case for most materials. In practice stirring effects are also present in the weld pool. These effects can be modelled by assigning an enhanced value to the thermal conductivity of the molten material. This can result in an order of magnitude change in thermal conductivity around the melting temperature which could prove to be a potential source of trouble in any numerical scheme. This problem is overcome by working with a transformation of the conductivity variable. The stirring rates in various regions of the weld pool can also be different. This is simulated by having different conductivity-temperature relationships for the various regions.

2. FORMULATION

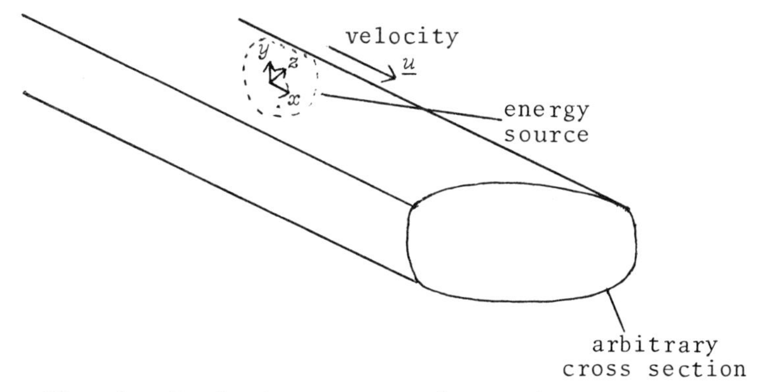

Fig. 1. Typical geometry of material strip.

The material geometry is assumed to take the form of an infinite strip with a constant arbitrary cross section with the energy source moving in a direction normal to this cross section, as illustrated in Fig. 1. With respect to a reference frame fixed in the energy source the energy conservation equation takes the form:

$$\frac{\partial H}{\partial t} + \underline{u}(t).\nabla H - \nabla.(k\nabla T) = Q(\underline{x},t) \qquad (2.1)$$

where H is the thermal energy or heat content of the material. A typical H,T relationship is shown in Fig. 2 where the rapid increase around the melting temperature is a result of the latent heat of the material. A typical k,T relationship is shown in Fig. 3 where the rapid increase around the melting temperature is to model the weld pool stirring. To eliminate numerical problems a conductivity potential ϕ is now introduced, following Albasiny [1] and Eyres *et al* [2], where

$$\phi = \int^{T} k \, dT \ . \qquad (2.2)$$

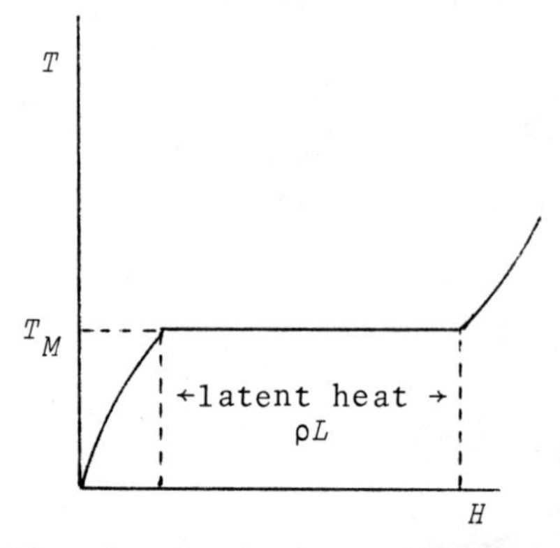

Fig. 2. Typical relationship between thermal energy H and temperature T.

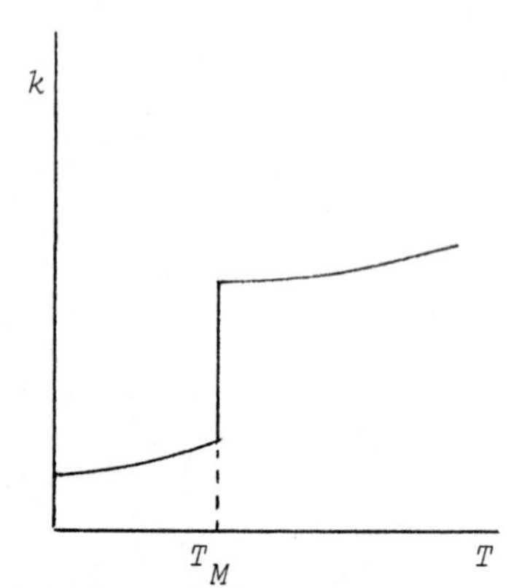

Fig. 3. Typical relationship between thermal conductivity k and temperature T.

The energy equation now becomes:

$$\frac{\partial H}{\partial t} + \underline{u}.(\nabla H) - \nabla^2\phi = Q(\underline{x},t) . \qquad (2.3)$$

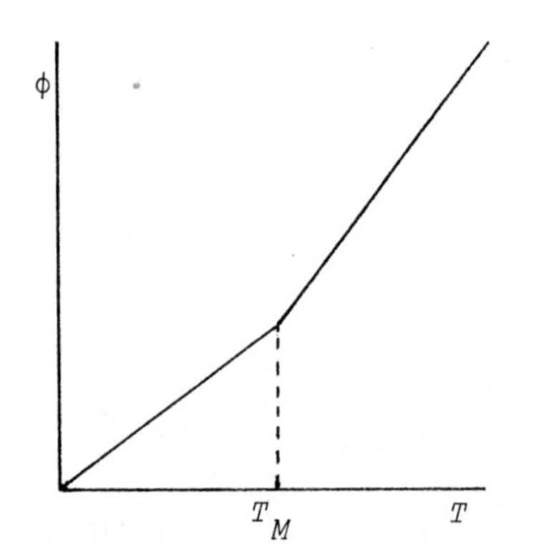

Fig. 4. Typical relationship between conductivity potential ϕ and temperature T.

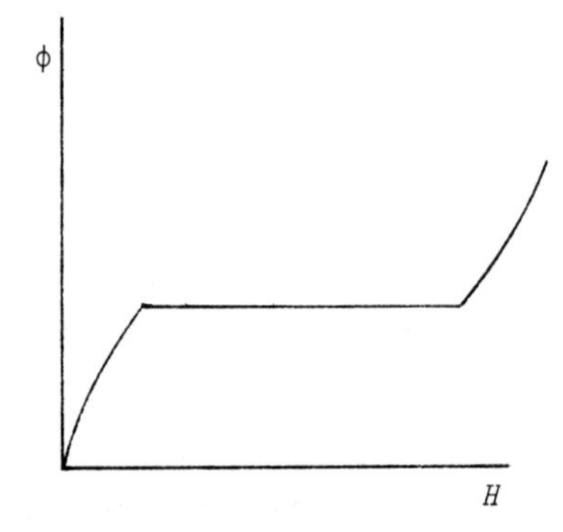

Fig. 5. Typical relationship between conductivity potential ϕ and thermal energy H.

The introduction of the variable ϕ gives a ϕ,T relationship of the form shown in Fig. 4 and a ϕ,H relationship of the form shown in Fig. 5. From this it can be seen that the use of ϕ as the independent variable in a numerical scheme could give trouble around the melting temperature where small changes in

ϕ produce large changes in H. For this reason in the following scheme H will be used for the main independent variable with ϕ regarded as a function of H. Thus (2.3) is to be solved for H and then the corresponding temperature distribution obtained from the known T,H relationship.

3. NUMERICAL METHOD

3.1 Finite Difference Equation

The system is divided up into volume cells, fixed with respect to the reference frame, which are formed from sub-divisions of cross-sectional segments of the infinite strip of material, with each segment subdivided in the same manner, as illustrated in Fig. 6.

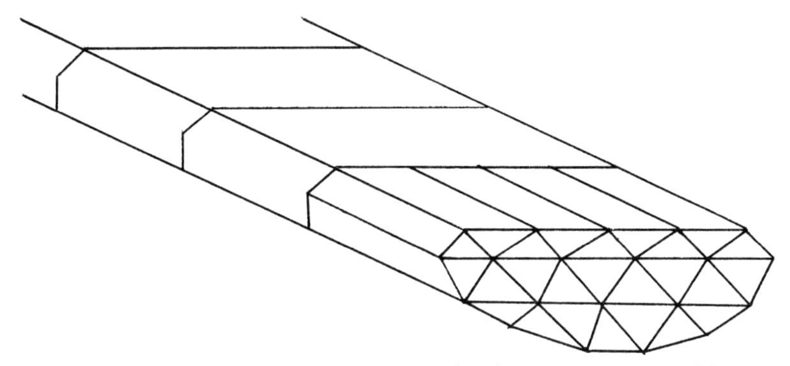

Fig. 6. Subdivision of material strip into volume cells.

A mean thermal energy is defined for each cell by:

$$H_i = \frac{1}{V_i} \int_{V_i} H \, dV \tag{3.1}$$

where V_i is the volume of cell number i.

Integrating (2.3) over the volume V_i gives

$$\frac{\partial H_i}{\partial t} = Q_i - \int_{S_i} H \, \underline{u} . d\underline{S} + \int_{S_i} \nabla\phi . d\underline{S} \tag{3.2}$$

where $Q_i = \frac{1}{V_i} \int_{V_i} Q \, dV$ is the mean energy input rate into cell i.

To reduce (3.2) to a finite-difference form an expression is required for the surface integrals in terms of H_i, $\phi_i (=\phi(H_i))$, and the corresponding values of H_j, ϕ_j for the cells j adjacent to cell i. For cells on the boundary the appropriate boundary condition is incorporated in the representation of the surface integral. This results in a set of nonlinear spatial finite-difference equations of the form:

$$\frac{\partial \underline{H}}{\partial t} = \underline{F}(\underline{H}) \tag{3.3}$$

where $\underline{H}$ is a vector with components H_i and $\underline{F}$ is a nonlinear function whose ith component F_i is a function of H_i and the values H_j for the cells j adjacent to cell i.

3.2 Representation of Surface Integrals

The components of the first surface integral in (3.2) represent a convection energy flux and are identically zero for all parts of cell surfaces except for those normal to $\underline{u}$. For these surfaces, which lie on cross-sections of the infinite strip, a value of H can be obtained from a first order Taylor expansion resulting in:

$$\int_{S_{ij}} H\underline{u}.d\underline{S} = \left(\frac{d_i H_j + d_j H_i}{d_i + d_j}\right) \underline{u}.\underline{n}_{ij} S_{ij} \tag{3.4}$$

where S_{ij} is the common surface of adjacent cells i and j
$\underline{n}_{ij}$ is the unit normal from cell i into cell j
d_i, d_j are perpendicular distances from cell centres to S_{ij}.

The second surface integral in (3.2) can likewise be obtained from a first order Taylor expansion which is equivalent to an assumption of a local steady state. This gives:

$$\int_{S_{ij}} \nabla\phi.d\underline{S} = \left(\frac{\phi_j - \phi_i}{d_i + d_j}\right) S_{ij} \;. \tag{3.5}$$

However this expression is only valid for adjacent cells with the same k,T relationship. For adjacent cells with different k,T relationships the requirement for temperature continuity would produce a discontinuity in ϕ at the common surface. The evaluation of an expression for $\nabla\phi$ would then involve the solution of integral equations expressing temperature and flux continuity at the common surface. Since this would re-introduce thermal conductivity terms directly an alternative approach is indicated. The difficulty can be overcome if the boundaries between regions with different k,T relationships pass through the cell centres in such a way that the sections between adjacent cell centres have the same k,T relationship. The simplest way to achieve this is to restrict regions with the same k,T relationship to complete cross sections of the infinite strip with region boundaries lying in the middle of cross sectional segments, as shown in Fig. 7. With this restriction expression (3.5) can be used for the various parts of the surface of cell i where ϕ_i denotes $\phi(H_i)$ evaluated for the material of the region containing the surface S_{ij}.

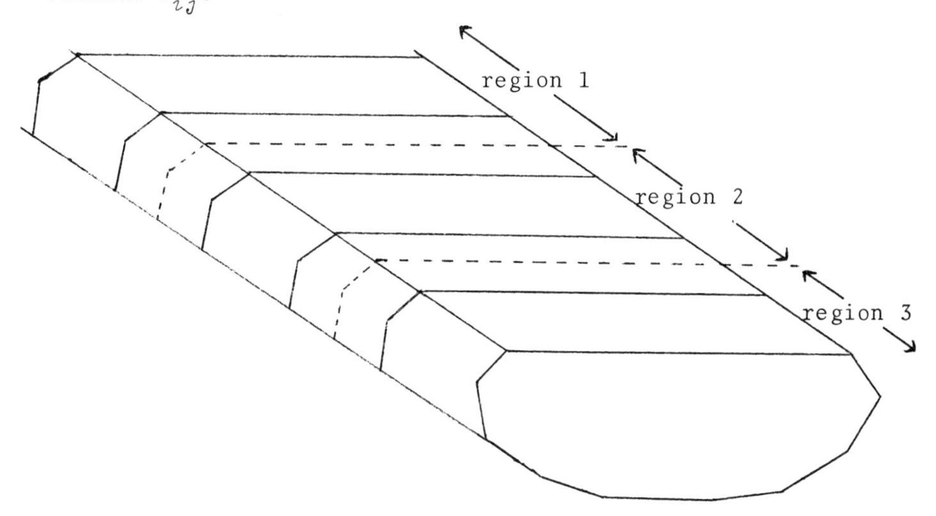

Fig. 7. Subdivision of material strip into regions with different conductivity-temperature relationships.

3.3 Solution of Difference Equation

In order to obtain the solution to the nonlinear equation (3.3), with an initial thermal energy distribution $\underline{H}^o$, a time step approach will be used. To update the solution from time t^n to time t^{n+1} an implicit Crank-Nicolson type scheme is used:

$$\underline{H}^{n+1} = \underline{H}^n + \frac{\delta t}{2}\left[\left(\frac{\partial \underline{H}}{\partial t}\right)^n + \left(\frac{\partial \underline{H}}{\partial t}\right)^{n+1}\right]$$

$$= \underline{H}^n + \frac{\delta t}{2}\left[\underline{F}(\underline{H}^n) + \underline{F}(\underline{H}^{n+1})\right] \quad . \tag{3.6}$$

An iterative method is required to solve this nonlinear equation for $\underline{H}^{n+1}$. One fairly powerful scheme uses Newton's method. An initial estimate for $\underline{H}^{n+1}$ is given by:

$$\underline{H}_0^{n+1} = \underline{H}^n + \delta t \; \underline{F}(\underline{H}^n) \; . \tag{3.7}$$

The initial estimate is them improved by a sequence of corrections $\underline{h}_m$ which are obtained from:

$$(I - J)\underline{h}_m = \underline{H}^n + \frac{\delta t}{2}\underline{F}(\underline{H}^n) - \underline{H}_m^{n+1} + \frac{\delta t}{2}\underline{F}(\underline{H}_m^{n+1}) \tag{3.8}$$

where

$$\underline{H}_m^{n+1} = \underline{H}_0^{n+1} + \sum_{i=0}^{m-1} \underline{h}_i \quad , \tag{3.9}$$

I is the identity matrix,

J is a Jacobian matrix with components given by

$$J_{ij} = \frac{\delta t}{2}\left(\frac{\partial F_i}{\partial H_j}\right)\Bigg|_{\underline{H} = \underline{H}_m^{n+1}} \; . \tag{3.10}$$

The solution of (3.8) for the correction $\underline{h}_m$ involves the inversion of the banded matrix $(I - J)$. The same inverse can be used in the calculation of several successive corrections to speed up the computational process.

For many problems the computer storage required for the inversion of the matrix $(I - J)$ can prove excessive. In these circumstances an alternative method for the solution of (3.6) is to use nonlinear successive over-relaxation. Here the method proceeds as before except that the components of the mth correction $\underline{h}_m$ are now obtained from:

$$(\underline{h}_m)_i = w\frac{\left[\underline{H}^n + \frac{\delta t}{2}\underline{F}(\underline{H}^n) - \underline{H}_m^{n+1} + \frac{\delta t}{2}\underline{F}(\underline{H}_m^{n+1})\right]_i}{(1 - J_{ii})} \tag{3.11}$$

where w is a relaxation parameter.

REFERENCES

1. Albasiny, E.L., *Proc. IEE*, **103**, Series B, 158 (1956).

2. Eyres, N.R., Hartree, D.R., Ingham, J., Jackson, R., Sarjant, R.J. and Wagstaff, J.B., *Phil. Trans. Roy. Soc. of London*, **A**, **240**, 1 (1946).

Chemical and Biological Problems

John Crank
(Brunel University)

1. STEFAN-TYPE DIFFUSION PROBLEMS

A general class of problem relates to the diffusion of particles into a medium which can trap some of them and prevent them from taking further part in the diffusion process. If we like, we can think of the medium as containing a number of holes in which some diffusing particles are caught. We are interested in immobilisation processes which are irreversible and so rapid compared with the rate of diffusion that they may be considered instantaneous. Only a limited number of molecules can be trapped in a given volume of the medium.

The trapping may be due to a chemical reaction by which the diffusing molecules are either precipitated or form a new immobile chemical compound. Alternatively, we may have a less specific physical adsorption onto fixed attractive sites of some kind. An example of biological interest is the diffusion of oxygen into muscle when the oxygen combines with lactic acid. In the textile industry we have the process of dyeing a fabric with certain dyes, and the reaction of copper ions with CS_2 groups as they diffuse into cellulose xanthate during the spinning of cellulose fibres.

In all these cases a sharp boundary moves through the medium. It separates a region in which all the holes or sites are occupied from one in which none of them are. In front of the advancing boundary the concentration of freely diffusing molecules is zero, while behind it immobilisation is complete. Fig. 1 illustrates the situation.

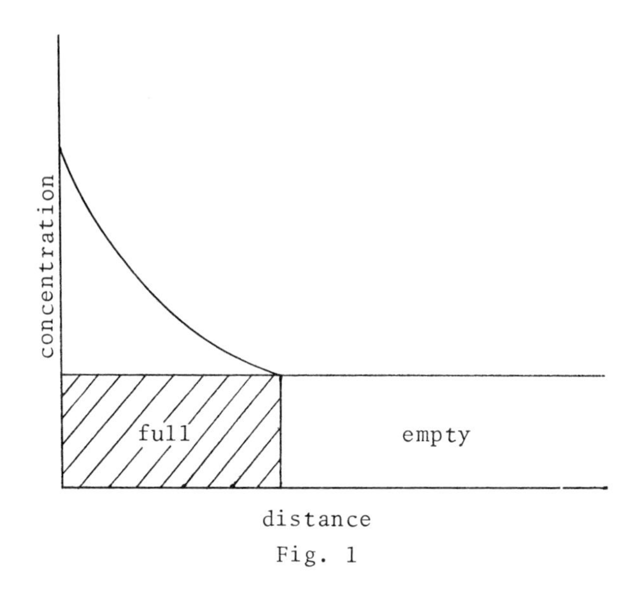

Fig. 1

Other moving boundary problems are common in diffusion. A film of tarnish forms on the surface of a metal by reaction with a gas which diffuses through the film of progressively increasing thickness [5]. A corresponding problem with radial symmetry relates to the diffusion-controlled growth of a new phase such as a crystal in solution [6]. When evaporation occurs from the free surface of a chemical solution the surface recedes and activates a back-diffusion of the solute into the remaining solution [16]. As an example in the field of soil mechanics we may refer to the progress of consolidation in a clay layer which increases in thickness with time [10].

2. DIFFUSION: CHARACTERISTIC FEATURES

While formally the mathematics of heat flow and diffusion are the same, two important properties of practical systems tend to be different for the two phenomena. While the coefficient of heat conductivity can vary with temperature usually the extent of the variation is much less than that of the corresponding diffusion coefficient. The latter can vary by several hundredfold or more in one experiment. It frequently increases linearly or exponentially with increasing concentration and so the need for numerical solutions, rather than analytic solutions based on the assumption of a constant diffusion coefficient, is more acute.

Secondly, the immobilising reaction is quite likely to cause a volume change on one side or other of the moving boundary on which it takes place. Again this often leads to more significant relative motion of the media on the two sides of the boundary than happens in the analogous phase change problems in heat flow. Generalised mathematical formulations which allow for relative motion of the two media have been developed by Danckwerts [5].

3. DISCONTINUOUS DIFFUSION COEFFICIENTS

A diffusion coefficient which changes discontinuously from one value to another at a prescribed concentration gives rise to a discontinuous change in concentration gradient and hence a moving boundary. In the extreme case of a diffusion coefficient which is zero in the range $0 \leqslant C \leqslant C_s$ and has a constant finite value D_1 for $C_s \leqslant C \leqslant C_1$ as in Fig. 2a a discontinuous change in the concentration itself occurs as in Fig. 2b.

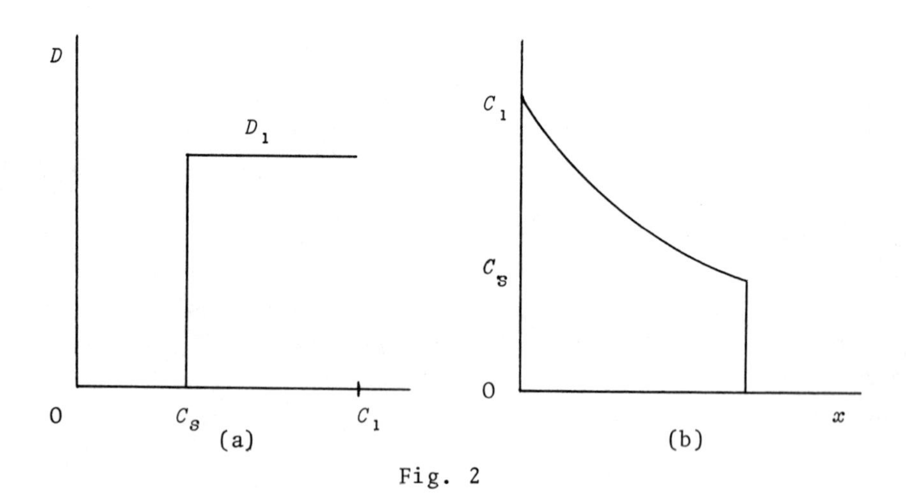

Fig. 2

These plots have the form of Fig. 1. If we interpret C to be the total concentration of molecules including both freely diffusing and immobilised, then C_s can be interpreted as the saturation concentration of the immobilised component: it is only for values of C greater than C_s that diffusion is possible. Thus a problem in diffusion with immobilisation can be thought of formally as a concentration-dependent diffusion process governed by a diffusion coefficient which changes discontinuously at one concentration. The mathematical solution is the same for both problems.

4. BIOLOGICAL DIFFUSION PROBLEMS

The diffusion of oxygen in tissue which simultaneously consumes oxygen occurs in a number of biochemical and medical situations. One such situation arises in research into the irradiation of tumour tissue. The killing effectiveness of the radiation on cancerous cells depends, among other factors, on the oxygen content of the cells. The need to interpret experimental measurements has led to a mathematical formulation and solution of the problem of oxygen diffusing into and out of a plane sheet of a medium which absorbs oxygen at a constant rate [4]. A moving boundary marks the innermost limit of oxygen penetration. The mathematical solutions allow the experimentalist to relate the results of measurements made on the surface of the medium to the rate of oxygen consumption within the medium.

Stated in its simplest, non-dimensional form the problem requires solutions of the following system:

$$\frac{\partial c}{\partial t} = \frac{\partial^2 c}{\partial x^2} - 1, \quad 0 \leqslant x \leqslant x_0(t) \tag{4.1a}$$

$$\frac{\partial c}{\partial x} = 0, \quad x = 0, \quad t \geqslant 0 \tag{4.1b}$$

$$c = \partial c/\partial x = 0, \quad x = x_0(t), \quad t \geqslant 0 \tag{4.1c}$$

$$c = \tfrac{1}{2}(1 - x)^2, \quad 0 \leqslant x \leqslant 1, \quad t = 0. \tag{4.1d}$$

The formulation of these equations and their solution by finite-difference methods will be described later (pp. 198-200)*.

* See also Tayler (p. 125) and Ferriss (p. 251).

Crank and Gupta [4] combined a numerical solution with an integral or heat-balance method [11] in order to obtain an approximate analytic solution of the problem.

In the range $0 \leqslant t \leqslant 0.020$ the boundary has not moved appreciably and the expression:

$$c(x,t) = \tfrac{1}{2}(1-x)^2 - 2(t/\pi)^{\frac{1}{2}} \exp\left[-(\tfrac{1}{2}x/t^{\frac{1}{2}})^2\right] + x \operatorname{erfc}(x/2t^{\frac{1}{2}}) \quad (4.2)$$

is sufficiently accurate for most practical purposes. Numerical solutions confirm that the concentration on the surface $x=0$ varies linearly with the square root of time and is given by:

$$c(0,t) = \tfrac{1}{2} - 2(t/\pi)^{\frac{1}{2}} = c_0 \text{ say.} \quad (4.3)$$

A polynomial profile of the fourth degree suitable for the application of Goodman's method leads to:

$$c(x,x_0) = (1-x/x_0)^2\left[\tfrac{1}{2}x^2+4c_0(1-x/x_0)-3c_0(1-x/x_0)^2\right] \quad (4.4)$$

where $x_0(t)$, the position of the moving boundary, is the solution of the ordinary differential equation:

$$\frac{dx_0}{dt} = -\frac{(20 - 8/(\pi t)^{\frac{1}{2}})x_0}{x_0^2 + 4 - 16(t/\pi)^{\frac{1}{2}}} \quad (4.5)$$

The Runge-Kutta numerical solution for $x_0(t)$ can be approximated by the expression:

$$x_0(t) = 1 - \exp\left[-2(t_1 - t)^{\frac{1}{2}}/(t - t_0)^{\frac{1}{2}}\right] \quad (4.6)$$

where $t_0 = 4/25\pi$ and is the minimum time for which the differential equation for $x_0(t)$ is valid, since we know $dx_0/dt \leqslant 0$ and $t_1 = \pi/16$, which from (4.3) is the time for the surface concentration, $c(0,t)$, to reach zero.

Thus we now have analytical expressions describing the movement of the boundary and the concentration profile for all times.

Closely similar mathematical problems are posed by attempts to distinguish between inhibitor or stimulator mechanisms of cell growth [3].

5. ARTIFICIAL MOVING BOUNDARIES

It is a fundamental property of mathematical solutions of

the heat flow or diffusion equation in infinite or semi-infinite media that a disturbance on the surface is propagated instantaneously to an infinite distance below the surface. Any specified concentration of finite magnitude, however, penetrates with a finite speed, the distance often being proportional to the square root of time. To any given degree of accuracy, therefore, an approximate mathematical solution can incorporate the idea of a penetrating boundary which marks a small but finite concentration. This is a mathematical artefact rather than a physical sharp boundary, but is a useful way of developing approximate analytic or numerical solutions.

6. DIFFUSION ANOMALIES IN POLYMERS

When a liquid penetrant diffuses into a polymer sheet or filament, sharp boundaries can often be seen under a microscope. The topic is reviewed by Park [18] and Rogers [20]. Hartley [12] noted three boundaries. The inner boundary marks the limit of penetration of the liquid, while the outer boundary shows the outer edge of the swollen gel. The third intermediate boundary is usually thought to lie between polymer in the elastic rubbery state and glassy polymer. The essential distinction is that polymers in the rubbery state respond rapidly to changes in their condition. For example, a change in temperature causes an almost immediate change to a new equilibrium volume. The properties of a glassy polymer, however, tend to be time-dependent. For example, stresses may be slow to decay after such a polymer has been stretched.

Often the positions of the boundaries vary according to $(\text{time})^{\frac{1}{2}}$. But boundaries moving at a constant speed have been observed and some follow intermediate power laws (Hartley [12], Kwei and Zupko [15]). Wang *et al* [21] suggest that the depth of penetration can be expressed as $at + bt^2$ where t is time and a and b are constants.

Interferometric fringe photographs of chloroform penetrating cellulose acetate, for example, between parallel glass plates show that the concentration of chloroform changes very rapidly near the inner boundary [19]. This observation suggests the existence of large internal stresses between the swelling polymer and the central unattacked core of the sheet. This view is substantiated by other photographs which show transverse cracks in the unswollen core and by evidence of anisotropic dimensional changes [13] and [1]. For example, the radial swelling of a fibre is not at first accompanied by an increase in fibre length, which only starts when the sharp boundary reaches the centre of the fibre.

Thus we are here concerned with moving boundary problems in which the diffusion process is complicated by the fact that part of the polymer medium may be slowly relaxing and changing its properties at a rate dependent on the concentration of the penetrant molecules. There may also be time-dependent internal stresses, differing from point to point, which affect the diffusion process. Finally, the geometry of the polymer sheet is changing with time as it swells, probably anisotropically.

A number of mathematical models of these penetrant-polymer diffusion phenomena have been constructed [2], [7], [8], [9], [17] and [21].

Alfrey *et al* [1] proposed a useful classification according to the relative rates of diffusion and polymer relaxation. Three classes are distinguished:

(*a*) Case 1 or Fickian diffusion in which the rate of diffusion is much less than that of relaxation.

(*b*) Case 2 is the other extreme in which diffusion is very rapid compared with the relaxation process.

(*c*) Non-Fickian or anomalous diffusion which occurs when the diffusion and relaxation rates are comparable.

The significance of the terms Case 1 and Case 2 is that they are both simple cases in the sense that the behaviour of each can be described in terms of a single parameter. Case 1 systems are controlled by a diffusion coefficient. In Case 2 the parameter is the constant velocity of the boundary between swollen gel and glassy polymer.

Crank [2] examined a simple model based on the assumption that a compressive force decreases the diffusion coefficient below that for the unstressed material and vice versa for an extensive force. As diffusion proceeds a sharp advancing boundary separates a swollen region under compression from an unswollen inner core under extension. This general idea has since been extended by Frisch *et al* [9], Wang *et al* [21] and Kwei *et al* [14], who include in the basic transport equation, in addition to the gradient of chemical potential, a second term deriving from the partial stress of the penetrant. They deduce a generalised diffusion equation:

$$\frac{\partial c}{\partial t} = \frac{\partial}{\partial x}\left[D\frac{\partial c}{\partial x} - vc\right] \tag{6.1}$$

where v represents the velocity of the penetrating molecules attributable to the internal stresses. By coupling this with a diffusion coefficient which drops discontinuously to zero at some concentration, many of the features of the behaviour of these swelling polymers can be described semi-quantitatively.

No single model successfully predicts all the experimental observations. Each one accounts for some of the salient features and collectively they contribute towards an elucidation, albeit incomplete, of the complex mechanisms involved. A more complex model is still needed to synthesise the effects of polymer relaxation, orientation and stress on the diffusing swelling process.

Even more complicated two-dimensional problems are posed at the corner of a swelling sheet in which the polymer molecules have initially a preferential direction of orientation.

REFERENCES

1. Alfrey, A., Gurnee, E.F. and Lloyd, W.G., *J. Poly. Sci.* **C12**, 249 (1966).
2. Crank, J., *J. Poly. Sci.* **11**, 151 (1953).
3. Crank, J., *Bull. Inst. Maths. Applics.* Vol. **10**, Nos. 1 and 2 (1974).
4. Crank, J. and Gupta, R.S., *J. Inst. Maths. Applics.* **10**, 19 and 296 (1972).

5. Danckwerts, P.V., *Trans. Faraday Soc.* 46, 701 (1950).

6. Frank, F.C., *Proc. Roy. Soc.* 201A, 586 (1950).

7. Frisch, H.L., *J. Chem. Phys.* 41, 3679 (1964).

8. Frisch, H.L., in "Non-Equilibrium Thermodynamics Variational Techniques and Stability" *editors* Donnelly, R.J., Herman, R. and Prigogine, I., p. 277, University of Chicago (1966).

9. Frisch, H.L., Wang, T.T. and Kwei, T.K., *J. Poly. Sci.* A2, 7, 879 (1969).

10. Gibson, R.E., *Quart. Appl. Math.* 18, 123 (1960).

11. Goodman, T.R., *ASME Transactions* 80, 335 (1958).

12. Hartley, G.S., *Trans. Faraday Soc.* 42B, 6 (1946).

13. Hermans, P.H., "A contribution to the Physics of Cellulose Fibres," p. 23, Elsevier, Amsterdam (1948).

14. Kwei, T.K., Wang, T.T. and Zupko, H.M., *Macromolecules* 5, 645 (1972).

15. Kwei, T.K. and Zupko, H.M., *J. Poly. Sci.* A2, 7, 867 (1969).

16. Meadley, C.K., *Quart. J. Mech. and Appl. Math.* 24, 43 (1971).

17. Newns, A.C., *Trans. Faraday Soc.* 52, 1533 (1956).

18. Park, G.S., in "Diffusion in Polymers," *editors* Crank, J. and Park, G.S., Chap. 5, Academic Press (1968).

19. Robinson, C., *Proc. Roy. Soc.* 204A, 339 (1950).

20. Rogers, C.E., in "Physics and Chemistry of the Organic Solid State," *editors* Fox, D., Sabes, M.M. and Weissberger, A Vol. II, Chap. 6, Interscience (1965).

21. Wang, T.T., Kwei, T.K. and Frisch, H.L., *J. Poly. Sci.* A2, 7, 2019 (1969).

CURRENT PROBLEMS IN THE GLASS INDUSTRY

D. Gelder and A.G. Guy
(Pilkington Brothers Research Laboratories)

1. INTRODUCTION

In considering moving boundary problems in the glass industry, and no doubt in other industries, the first essential in any particular case is to determine what is required of the analysis. An accurate solution of a well defined problem is rarely the main priority. In so far as there is a typical requirement, it is to determine the main factors controlling some phenomenon, and to compare the effectiveness of various changes which seem likely to give improved behaviour. In retrospect the use of numerical methods is rarely essential, but in practice a numerical simulation may seem preferable to justifying the approximations necessary for an analytical treatment. Two problems are examined in some detail below, with an emphasis on meeting the practical requirements rather than resolving the mathematical difficulties which arise. In both cases heat transfer rather than diffusion is involved, and the concept of a sharp interface is only a first approximation to the truth. There are many problems of equal importance involving diffusion, and a brief review of these is given here, although the lack of data on diffusion coefficients is generally such that a detailed mathematical treatment cannot be justified.

In glass melting, the solution of sand grains in the surrounding liquid is thought to be much the slowest of the various reactions involved, and a quantitative theoretical treatment would be of interest and perhaps of use. The converse problem of crystal growth arises below the "liquidus" temperature of a glass. In general the aim is to avoid crystals by cooling sufficiently rapidly through the temperatures at which the growth rate is significant. However the production of glass ceramics depends on close control of the crystal growth in circumstances where there may be significant heat evolution. In these problems the composition changes near the solid surface are liable to give large variations in diffusion coefficient from that for the bulk glass. The elimination of bubbles is another aspect of glass melting which is largely a diffusion controlled moving boundary problem. The general decrease in gas solubilities as temperature rises means that the larger bubbles grow and rise out of the melt, while smaller ones hardly move and disappear as the temperature drops from its peak value. The development of metallic inclusions (for example in solar control glasses), the behaviour of glasses which are liable to separate into distinct phases, and some aspects of the corrosion of refractories by glass are other problem areas involving moving interfaces. A general discussion of many of these problems appears in Pye *et al* [8], but in most of them neither fundamental theory nor experimental measurements are able to give sufficient physical data to permit detailed calculations.

The first problem to be considered in detail here is "batch melting", the basis of glass making. "Batch", consisting of sand, lime, soda ash and other minor constituents, is heated together with cold glass resulting from losses during processing and converted into molten glass. The thermal conductivity of the batch is low so that the interior is scarcely affected before the surface starts to melt, and the interface between molten glass and solid batch then moves through the batch, forming a moving boundary. The simplest view is that all the chemical reactions take place at some well defined temperature, say 850°C, with a known heat requirement. The batch is heated primarily from above, and although the major chemical reactions take place at around 850°C the sand does not dissolve readily in the resulting liquid until perhaps 1450°C. A rapid increase in thermal conductivity with temperature, and uncertainty as to the actual values of conductivity, are the only complicating factors on this view of the process, and a satisfactory numerical investigation will be described. However for some purposes it is necessary to take into account the non-zero thickness of the melting zone, and the treatment of this and other refinements is still incomplete.

The second problem concerns the use of certain electrically conducting glasses as switches. Various mechanisms are thought to be involved, depending upon the glass composition used. "Thermal switching" is examined here, in circumstances where the glass undergoes a phase transition with a substantial increase in electrical conductivity at perhaps 70°C. On putting an increasing voltage across such a glass, at some critical value Joule heating will raise the temperature sufficiently for a highly conducting core to develop. The outside of this core forms a moving boundary, and interest normally centres on periodic operation. The effect of the internal heating is to give a finite width transition zone as the core first expands and a sharp change later as it contracts. This variation in the nature of the moving boundary complicates the use of some of the special numerical techniques for these problems.

2. BATCH MELTING

On large flat glass tanks the batch is fed as a blanket which is pushed forward on to the tank. The density is much less than that of glass so that it floats on the surface. Radiant heating is supplied from above by flames, and some heat comes also from the glass beneath. As the blanket moves forward, a surface layer of gradually increasing depth is converted into glass. The general situation is illustrated in Fig. 1. On small tanks the heat may be supplied entirely by electrical resistance heating of the glass. The batch is then fed from above, and all the heat to melt it comes from the glass beneath. Another method of feeding involves comparatively narrow ribbons rather than a broad blanket being pushed forward. Particularly with this last system, the hot surface glass may be of low enough viscosity to run off. Apart from this effect the melting may always be regarded as a one dimensional vertical process (since horizontal heat transfer is negligible), considering variations with time in a cylindrical section moving with the blanket in situations where the latter is being pushed forward.

Unfortunately the batch has a low thermal conductivity, and the basic thermal problem in melting is the transfer of heat from the surface to the interior of the blanket.

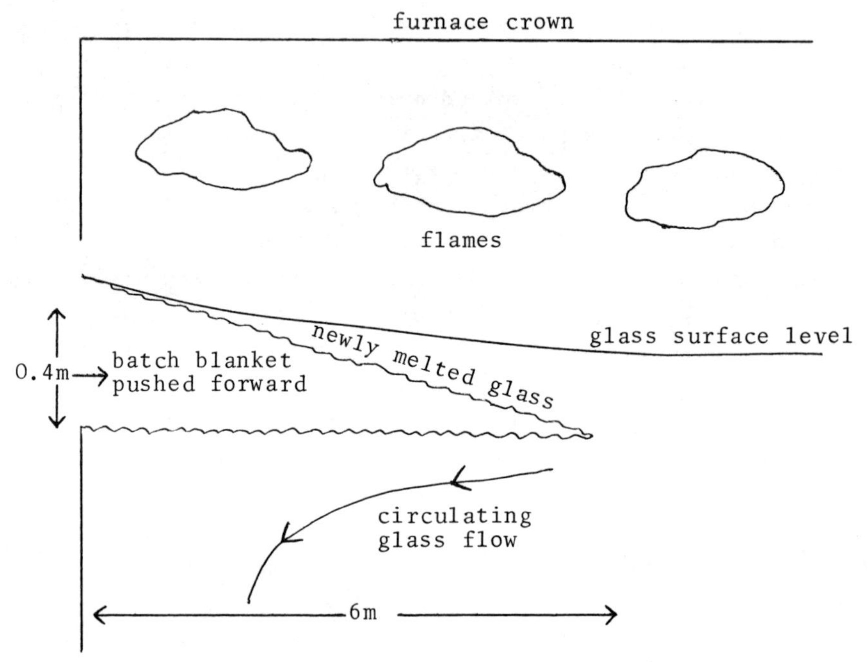

Fig. 1. Typical melting situation.

The rate of glass production per unit area of the blanket is an important measure of overall melting efficiency, and the primary aim of much theoretical and experimental work is to increase this rate of melting. In somewhat more detail, the requirements from theoretical studies are as follows:

(*i*) to distinguish the sort of changes from the present situation which would give substantial as opposed to marginal improvements;

(*ii*) to estimate the effects of the more obvious possible changes in operating conditions, and to be able to respond quickly with an analysis of any other moves which may be proposed;

(*iii*) to estimate the effects of changes in the batch which affect the chemistry of the process, and possibly also the final glass composition.

Pugh [7] gives a treatment along these lines, with a more detailed discussion of possible batch changes than will be attempted here. The moving boundary problem is solved by an analytical approximation, the most unsatisfactory feature being the neglect of the heat capacity of the batch above the reaction temperature. An analytical treatment is also possible taking a constant thermal conductivity above the reaction temperature, but this approach is still less satisfactory. Attention is here concentrated on the difficulties of producing a reliable mathematical model, and on the design of computer programs for the numerical solution of the resulting equations.

Consider first the history of a small volume of batch. Up to about 750°C it behaves as an inert material, and the thermal conductivity can be measured without difficulty: it tends to increase somewhat with temperature. The detailed variation depends on the composition and physical condition of the batch, but the conductivity is low compared with its value in later stages and can be supposed constant. At higher temperatures a complex chain of reactions occurs, the nett result being that by 900°C the batch may be regarded as glass with a large proportion of undissolved sand. The reactions require a substantial heat input, and as a first approximation they may be regarded as taking place at 850°C. At higher temperatures the sand dissolves (or strictly reacts), but this process is slow until 1450°C. Above 750°C the heat transfer properties of the batch are uncertain. The most extensive data are those of Kröger and Eligehausen [5] but the composition, grain size, atmosphere (and hence chemistry) and, most important, the time - temperature history in actual practice are liable to differ substantially from their experimental conditions. As the first liquid phases appear there is certainly a sudden jump in conductivity. However as the sand starts to disappear radiation can penetrate, and this probably becomes the main heat transfer mechanism. Large volumes of gas are liberated, and disturbances caused by this may lead to a nett vertical heat transfer. Radiation can be taken into account using the "effective conductivity" approximation, as described in Czerny and Genzel [4] based on absorption and scattering coefficients which can in principle be related to the bubble and residual sand content. The rate of dissolution of an individual sand grain could be calculated, but this is a moving boundary problem in itself, with very uncertain diffusion data. There is also the complication of gas rising through the melt, and an entirely theoretical approach to the heat transfer does not appear to have been attempted. Pugh [7] and the present work rely heavily on the data of Kröger and Eligehausen with the implicit assumption that the variations in the conductivity at a given temperature due to changes in temperature history are not significant.

A reasonable mathematical model for the melting of a vertical cylindrical section moving with the blanket is given below. In applying such a model the limitations imposed by overlooking much of what is important chemically must be remembered, but this is an aspect which will not be considered further here. Some care is needed in formulating the equations as there is a substantial change in density on melting, and the derivative $\frac{dT}{dt}$ is that following a batch particle. The following notation will be used:

T Temperature (°K for convenience in considering radiative transfer).

T_m "Melting temperature" - or reaction temperature.

L Heat of reaction.

K_b, K_g Thermal conductivity before and after reaction.

ρ_b, ρ_g Density before and after reaction.

C_b, C_g Specific heat before and after reaction.

y Vertical co-ordinate.

y_m, y_t Height above batch base of reaction level and top surface.

h, T_r Parameters giving radiative heat transfer from above.

Only K_g of the above physical properties is supposed to be temperature dependent. The model ignores the gas evolution, and the equations are as follows.

$$0 < y < y_m, \quad \frac{dT}{dt} = \frac{K_b}{\rho_b C_b} \frac{\partial^2 T}{\partial y^2}, \tag{2.1}$$

$$y_m < y < y_t, \quad \frac{dT}{dt} = \frac{1}{\rho_g C_g} \frac{\partial}{\partial y} \left(K_g \frac{\partial T}{\partial y}\right). \tag{2.2}$$

At $y = 0$, $\frac{\partial T}{\partial y} = 0$ (neglecting heat transfer from below). (2.3)

At $y = y_m$, $T = T_m$, (2.4)

$$\rho_b \frac{L dy_m}{dt} = \left(K_b \frac{\partial T}{\partial y}\right)_{y \to y_m -} - \left(K_g \frac{\partial T}{\partial y}\right)_{y \to y_m +}. \tag{2.5}$$

At $y = y_t$, $K_g \frac{\partial T}{\partial y} = h(T_r^4 - T^4)$, (2.6)

$$\frac{dy_t}{dt} = \left(1 - \frac{\rho_b}{\rho_g}\right) \frac{dy_m}{dt}. \tag{2.7}$$

At $t = 0$, $T = T_0$.

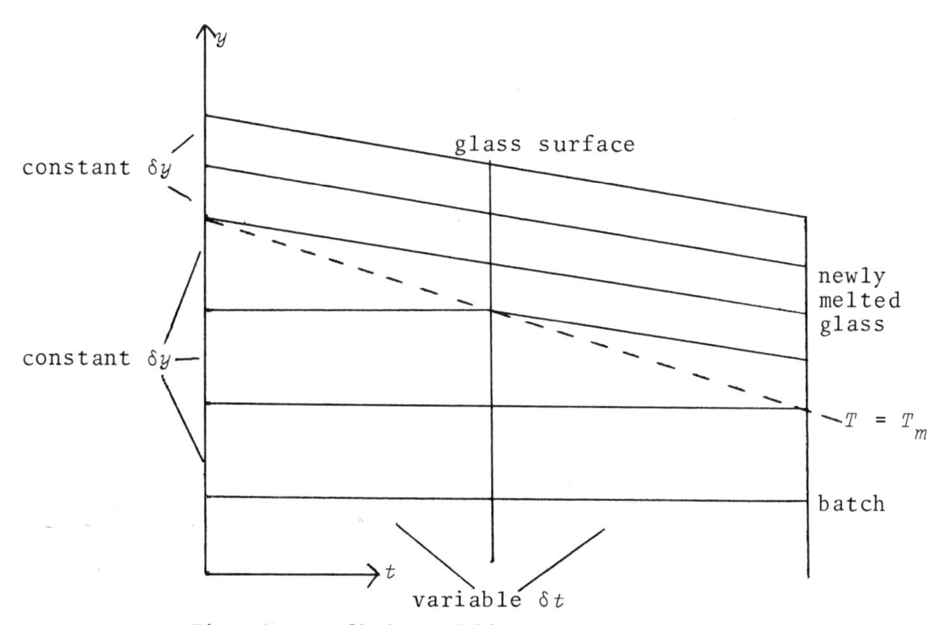

Fig. 2. A finite difference approach.

The main result required from (2.1) to (2.7) is the time t at which reaction is complete, *i.e.* when $y_m = 0$. K_g must be chosen taking account both of the experimental information and the need to obtain the correct result for present operating conditions. The model can then be used to predict the effect of various changes, whose overall benefit may sometimes be evaluated directly in terms of cost (*e.g.* raising T_0), or which may themselves involve considerable research (*e.g.* increasing K_g). The system can be solved in various ways, and one approach (illustrated in Fig. 2) has been to use a numerical technique with a fixed number of mesh points and a time step chosen so that y_m moves down by one mesh point.* It is of interest to note some of the details which prove important in obtaining a satisfactory computer program. The finite difference scheme used was entirely implicit, with simple one-sided differences for (2.3), (2.5) and (2.6) in the first instance. The first stage of the calculation, when $T < T_m$ everywhere, is straightforward, and the main calculation starts after interpolating to the situation when $T = T_m$ at the surface. A nonlinear system of equations is then found for T after the next step (since (2.6) is nonlinear and K_g depends on T). The time step itself is to be adjusted so that, solving the remaining equations, (2.5) is satisfied at the desired mesh point. These nonlinearities as they stand lead to a slow and unreliable calculation, but it proves possible to take advantage of special features of the problem so that the nonlinearity of (2.2) and (2.6) becomes irrelevant. For a trial time step, it may be supposed that (2.5) rather than (2.6) is satisfied. (2.1), (2.3) and (2.4) may be solved as a linear system, and using (2.5) an upward integration of (2.2) is possible. (2.6) then provides the corresponding T_r, and it is necessary only to adjust the time step so that T_r has the correct value. Some care is needed in programming this iteration, particularly at the first time step when a good estimate of the step is not available, but a reliable program can be obtained without undue difficulty. However a comparison of results at different mesh sizes revealed disappointing accuracy, and this was traced to the use of a simple first order difference for $\left(\frac{\partial T}{\partial y}\right)_{y \to y_m-}$ in (2.5).

Due to the comparatively low value of K_b, T falls from T_m to T_0 in a handful of mesh points with typical data, and even a second order difference yields little improvement. There may

* See Crank (p. 203).

be more elegant ways of overcoming this difficulty - for example a heat conserving difference approximation to (2.5) - but an entirely adequate technique proves to be an extrapolation for $\left(\frac{\partial T}{\partial y}\right)$ based on the exponential form of the approximate analytical solution in this region.

$$\left(\frac{\partial T}{\partial y}\right)_{y \to y_m -} = \frac{(T_i - T_{i-1})^2}{(T_{i-1} - T_{i-2})\delta y} \ . \qquad (2.8)$$

Here T_i, T_{i-1} and T_{i-2} are mesh point values, with T_i at the interface, *i.e.* $T_i = T_m$.

A heat balance method has also been used for this problem. It avoids the difficulties of obtaining a satisfactory discretisation, and gives results which agree well with those from the finite difference method. The overall effort required to obtain solutions by the two methods appears to be much the same.

The above approach provides a compact theory of batch melting, and it goes some way towards providing the necessary theoretical background. However, quite apart from doubts over regarding K_g as a known function of T, it leaves a good many questions unanswered. The effect of heat transfer from the glass beneath the blanket is of interest, and this introduces a second moving boundary. In addition it is of interest to consider the sudden increase in thermal conductivity as occurring at a different temperature from that at which the heat of reaction is taken in, and indeed to spread the latter over a temperature range. This serves partly as a check on the robustness of the simpler model, and also as an indication of the effect of batch changes which might influence these matters. To handle this range of requirements elegantly a computer program of considerable complexity is needed, and rather than develop the previous approach the other extreme of a smoothed-out program has been adopted. This is indeed a version of the Patankar-Spalding [6] parabolic equation program. The heat of reaction is specified as a large specific heat over an appropriate range of temperature. An important question is whether the difference equations should be treated as nonlinear in the physical properties at the end of a time step, or whether a linearised approach should be used. The latter was preferred, giving fast and reliable computing, but it is largely responsible for the need to use time steps around those for an explicit method to avoid unwelcome oscillations in results near the region of sharp change. The difference scheme is strictly conservative, but considerable attention to detail still proved necessary to obtain satisfactory performance. For example K_g is specified in terms of an absorption coefficient, and a mean value of this between mesh points (implying a harmonic mean K_g) gives disastrously slow melting rates with typical data. Using a mean value of K_g appears more reasonable, and is better both in comparing results at different mesh sizes and in comparison with the other program. After attention to other points of this nature which were only obvious in retrospect, the program has proved a satisfactory tool. As usual in calculations of this nature with rather poor accuracy, care is needed to ensure that results change in response to the data, and not to incidental changes in truncation error. The main practical implication appears to be that fixed time steps are preferable to automatically determined ones.

3. THERMAL SWITCHING

Certain glasses have electrical applications resulting from a sudden decrease in resistance which occurs when some critical voltage, which depends on the geometrical configuration, is applied. The mechanism is not the same in all glasses showing this effect, but it has become clear that in one instance the likely cause is a phase change occurring at quite a low temperature (about 70°C) and giving an increase in electrical conductivity by a factor of around 10^4. Boer [3] provides an extensive review for the case of a smooth but rapid increase with temperature. When a suitable voltage is applied in the cold state, sufficient current flows to give significant internal heating. As the voltage is slowly increased there comes a time when the transition temperature is reached at some point, and this leads to the development of a core of low resistance between the electrodes. At the same or even a considerably lower voltage this core would continue to grow, but the external circuit generally imposes a current limitation which keeps the core small. The phase change mechanism and the moving boundary problem for the core size are the subjects of the present study.

The requirements from theoretical studies are as follows:

(*i*) to confirm that this thermal switching mechanism does indeed explain the observed phenomena;

(*ii*) to determine the general characteristics of devices using this effect at both high and low frequencies of operation;

(*iii*) to indicate suitable device geometry for particular applications.

Berglund [2] provides an analysis for a film separated by a thin electrically insulating layer from a substrate of high thermal conductivity. Attention is here concentrated on situations where heat losses are primarily to more remote parts of the glass. The equations for periodic operation are also studied in more detail than by Berglund, although only a mathematical model involving severe approximations has been solved to date.

The following notation will be used:

T Temperature (°C).

T_0, T_t Ambient and transition temperatures.

L Latent heat for transition - this is very uncertain and may be negligibly small.

H, H_0 Enthalpy and value at T_t before transition.

σ_0, σ_t Electrical conductivity below and above T_t, both assumed to be constant for simplicity.

K Thermal conductivity.

ρ Density.

c Specific heat.

I, $\underline{i}$ Circuit current and current density vector in switch.

V, $\underline{E}$ Circuit voltage and electric field vector in switch.

R	Resistance in series with device.
ℓ	Electrode separation.
x, x_1	Co-ordinate across switch and half-width (where $T = T_0$).
d	Switch thickness.
r, r_1	Radial coordinate for axial symmetry and outer radius (where $T = T_0$).
s	Core size *i.e.* (x or r coordinate).
h	Heat transfer coefficient to substrate (where $T = T_0$).

The basic equations are:

$$\rho \frac{\partial H}{\partial t} = \nabla.(K\nabla T) + \frac{(\underline{i}.\underline{i})}{\sigma} \tag{3.1}$$

$$\underline{i} = \sigma \underline{E} \tag{3.2}$$

$$\nabla.\underline{i} = 0 \quad . \tag{3.3}$$

Here $\sigma = \sigma_0$ for $T < T_t$, $\sigma = \sigma_t$ for $T > T_t$ and σ is assumed to vary linearly between these values with H for $T = T_t$ as the latent heat is taken in or released. For the situations considered to date with plane geometry, d is small compared with the other dimensions, and T can be regarded as a function of x and t only (see Fig. 3). In this situation the heat loss to the substrate may be incorporated in (3.1), giving:

$$\rho \frac{\partial H}{\partial t} = \frac{\partial}{\partial x}\left(K \frac{\partial T}{\partial x}\right) + \frac{i^2}{\sigma} - \frac{h}{d}(T - T_0) \tag{3.4}$$

where the heat flow in the direction of the current flow has been ignored assuming that the heat flow through the electrodes is small. Before looking further at the time dependent equations, some steady solutions for specific geometries will be examined. These steady solutions give almost all the required information at low frequency operation.

Consider first plane geometry, with a thin film and heat losses primarily to the substrate. Berglund [2] considers this situation, and shows in particular that as K tends to zero the part of the core above T_t tends to an isothermal condition with temperature $2T_t - T_0$. When the switch is below T_t, the current density is given by:

$$i = \frac{\sigma_0 V}{\ell}$$

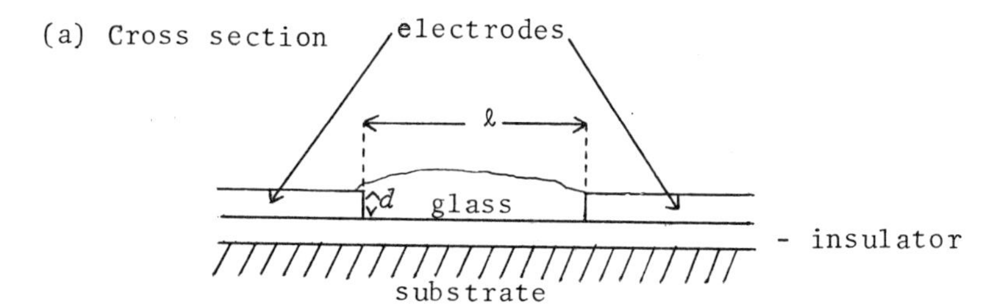

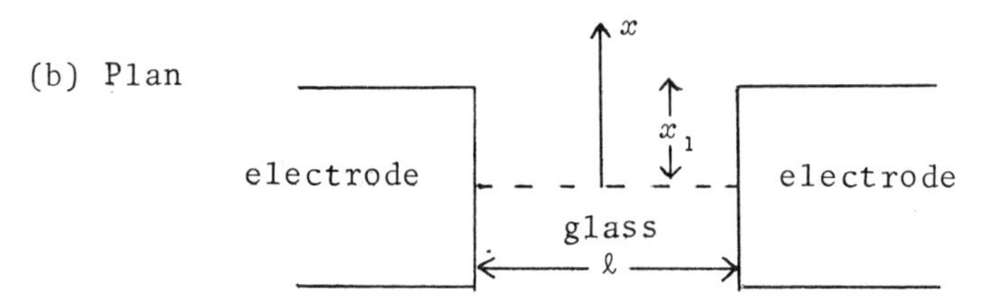

Fig. 3. General view of thin film switch.

This neglects R in comparison with $\ell/(2x_1 d\sigma_0)$. Switching will occur when:

$$\frac{i^2}{\sigma_0} = \frac{h}{d}(T_t - T_0)$$

i.e.
$$V^2 = \ell^2 h(T_t - T_0)/d\sigma_0 \tag{3.5}$$

Assuming for the moment that after switching the switch resistance is much less than R and neglecting the current carried in the high resistance region $x > s$, the new equilibrium state can be found easily. It is not unique for zero K, but for small K the conducting core will clearly be at the middle. For a core of half-width s, the current density is:

$$i = V/(2sdR) .$$

Using the result by Berglund that the core temperature is $2T_t - T_0$,

$$i^2 = \frac{2h\sigma_t}{d}(T_t - T_0)$$

i.e.
$$s^2 = V^2/8dhR^2\sigma_t(T_t - T_0) \quad . \tag{3.6}$$

This result applies for any V such that the simplifying assumptions are valid. At the transition value given by (3.4):

$$s^2 = \ell^2/8d^2R^2\sigma_t\sigma_0 \quad ,$$

i.e.
$$\frac{\ell}{2sd\sigma_t} = R\sqrt{\frac{2\sigma_0}{\sigma_t}} \quad . \tag{3.7}$$

Since $\sigma_t \sim 10^4 \sigma_0$, the switch resistance is indeed much less than R. The purpose of the resistance R is to prevent the core spreading to become the whole switch (partly to ensure rapid response, partly because the behaviour of the core is preferred for other reasons), and (3.6) implicitly assumes $s < x_1$. Since

$$\frac{s}{x_1} = \frac{\ell}{2x_1 d\sigma_0 R} \sqrt{\frac{\sigma_0}{2\sigma_t}} \tag{3.8}$$

it is clear that R must not be too small compared with the original switch resistance - a ratio of 1:14 gives $s \sim 0.1x_1$. The neglect of the current in $x > s$ is clearly satisfactory. Once switching has occurred, (3.6) shows that $s \propto V$ and thus $\underline{i}$ stays constant. However as V is decreased the switch resistance will become comparable with R. On the basis of (3.6) the two are equal when

$$V^2 = 2\ell^2 h (T_t - T_0)/d\sigma_t \tag{3.9}$$

and before V has fallen to this value the ratio s/V will be decreasing. It is clear by comparison with (3.5) that (3.9) gives in fact the minimum voltage to maintain a core, and this is referred to as the "holding voltage".

$$\frac{\text{holding voltage}}{\text{switching voltage}} = \sqrt{\frac{2\sigma_0}{\sigma_t}} .$$

Apart from the speed at which switching occurs, the above analysis gives a complete picture of the behaviour of one idealised switch for slow variations of V. However the heat transfer theory is a gross over-simplification of any practical situation and it is of interest to examine the same geometry but with the opposite extreme where lateral conduction predominates. Except for the ratio of power input in the hot core to that elsewhere (which is much greater here), this problem is mathematically identical with the welding problem described by Atthey [1]. The boundary conditions for the equation are then:

$$\text{at} \quad x = 0, \ \frac{\partial T}{\partial x} = 0; \quad \text{at} \quad x = x_1, \ T = T_0 .$$

When the switch is below T_t,

$$\frac{\sigma_0 V^2}{\ell^2} = -K \frac{\partial^2 T}{\partial x^2} \quad .$$

$$\text{At} \quad x = 0, \; T = T_0 + \frac{\sigma_0 V^2 x_1^{\,2}}{2\ell^2 K} \quad .$$

Again R has been neglected in comparison with $\ell/2x_1 d\sigma_0$. Switching will occur when:

$$V^2 = 2\ell^2 K \; (T_t - T_0)/x_1^{\,2}\sigma_0 \; . \tag{3.10}$$

With the previous assumptions after switching *i.e.* $\sigma_0 << \sigma_t$, and for a small core, the following boundary condition may be used:

$$\text{at} \quad x = s, \; \frac{\partial T}{\partial x} = \frac{T_0 - T_t}{x_1} \quad .$$

s is then given by the equation:

$$\frac{si^2}{\sigma_t} = \frac{K(T_t - T_0)}{x_1}$$

i.e. $$s = V^2 x_1/4d^2R^2\sigma_t K(T_t - T_0) \quad . \tag{3.11}$$

At the transition value given by (3.10):

$$s = \ell^2/2d^2R^2x_1\sigma_t\sigma_0$$

i.e. $$\frac{\ell}{2sd\sigma_t} \cdot \frac{\ell}{2x_1 d\sigma_0} = \frac{R^2}{2} \quad .$$

This differs considerably from the previous result, and it now follows immediately that if R is relatively small before switching, it is relatively large afterwards. An initial ratio of 1:14 would here give $s \sim 0.01x_1$. There is in this case an additional steady solution for V below the transition value in which the switch resistance is larger than R (but with the same power dissipation, so as to satisfy the boundary condition at $x = s$): however this is clearly unstable, and Fig. 4 shows it will not be reached in normal operation. Once switching has occurred, (3.11) shows that $s \propto V^2$, and the voltage over the switch is $1/V$ - it is inversely proportional to the total current. The switching off mechanism also differs. Instead of s decreasing to zero, switching occurs when $R = \ell/2sd\sigma_t$, with the maximum power dissipation in the switch for given V. The holding voltage is then given by:

$$V^2 = 8R\ell dK\ (T_t - T_0)/x_1 \tag{3.12}$$

$$\frac{\text{holding voltage}}{\text{switching voltage}} = \sqrt{2R/\frac{\ell}{2dx_1\sigma_0}} \quad .$$

At practical values of R the holding voltage is much nearer the switching voltage than in the previous situation.

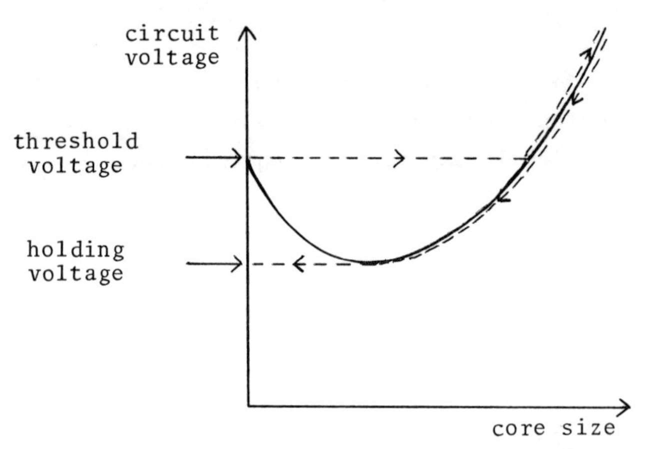

Fig. 4. Core size-voltage relation with heat loss through glass.

Clearly the properties of the switch depend to a considerable extent on the heat transfer situation. An alternative geometry for which a simple model is nearer the truth than in the above cases is axial symmetry. A one dimensional model with $T = T_0$ at $r = r_1$ appears quite reasonable. The result corresponding to (3.10) is:

$$V^2 = 4\ell^2 K\ (T_t - T_0) r_1{}^2 \sigma_0 \quad . \tag{3.13}$$

After switching, the appropriate boundary condition becomes:

$$\text{at} \quad r = s, \ \frac{\partial T}{\partial r} = \frac{T_0 - T_t}{s \log (r_1/s)} \quad .$$

s is given by the equation

$$\frac{\pi s^2 i^2}{\sigma_t} = \frac{2\pi s K\ (T_t - T_0)}{s \log (r_1/s)} \quad ,$$

$$i.e. \qquad s^2 = V^2 \log (r_1/s)/2R^2\sigma_t\pi^2 K\ (T_t - T_0) \ . \tag{3.14}$$

At the transition value given by (3.13):

$$s^2 = 2\ell^2 \log (r_1/s)/R^2 r_1{}^2 \sigma_t \sigma_0 \pi^2 \ ,$$

i.e. $$\frac{\ell}{\pi s^2 \sigma_t} \cdot \frac{\ell}{\pi r_1^2 \sigma_0} = \frac{R^2}{2 \log (r_1/s)} \quad .$$

The general behaviour is similar to the corresponding plane problem, although the results cannot be expressed as conveniently.

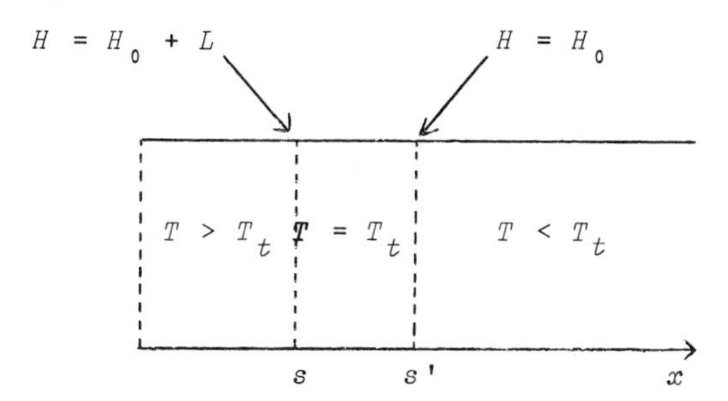

Fig. 5. General situation in time dependent operation

Considering now the time dependent problem, the general situation for the plane case with predominantly lateral heat loss is illustrated in Fig. 5. It is assumed that $\sigma_t > \sigma_0$ and $\frac{\partial T}{\partial x} \leqslant 0$ for $x \geqslant 0$ initially, and thus for all $t > 0$.

The electric field across the switch depends on the conductivity:

$$E = \frac{V}{\ell + 2R\int_0^{x_1} (\sigma d)\, dx} \quad .$$

In the three regions of Fig. 5, (3.1) takes the following forms.

$$x < s: \quad \rho c \frac{\partial T}{\partial t} = K \frac{\partial^2 T}{\partial x^2} + \sigma_t E^2 \tag{3.15}$$

$$x > s,\ x < s': \quad \rho \frac{\partial H}{\partial t} = \left\{ \sigma_0 + \frac{(H-H_0)(\sigma_t - \sigma_0)}{L} \right\} E^2 \tag{3.16}$$

$$x > s': \quad \rho c \frac{\partial T}{\partial t} = K \frac{\partial^2 T}{\partial x^2} + \sigma_0 E^2 \quad . \tag{3.17}$$

At $x = 0$, $\frac{\partial T}{\partial x} = 0$: at $x = x_1$, $T = T_0$.

It was assumed in the steady solution that $s = s'$, and it is clear from (3.16) that this is the case. When $s = s'$ in the time dependent problem, the usual condition holds for the movement of s:

$$L\frac{ds}{dt} = K\left(\frac{\partial T}{\partial x}\right)_{x\to s^+} - K\left(\frac{\partial T}{\partial x}\right)_{x\to s^-} . \tag{3.18}$$

Otherwise at $x = s$:

$$H_+ = H_{x\to s^+} < H_0 + L$$

$$\rho(L + H_0 - H_+)\frac{ds}{dt} = -K\left(\frac{\partial T}{\partial x}\right)_{x\to s^-}$$

and
$$\frac{dH_+}{dt} = \frac{\sigma_+ E^2}{\rho} + \frac{ds}{dt}\left(\frac{\partial H}{\partial x}\right)_{x\to s^+} .$$

These conditions are difficult to apply for small s. The final equation above gives a lower bound to the initial growth of s, since $\frac{dH_+}{dt}$ cannot be positive. Taking $\frac{dH_+}{dt} = 0$ (which amounts to neglecting the effect of conduction in the region $x < s$) may be shown to give an order of magnitude estimate as well as a lower bound for $\frac{ds}{dt}$. This estimate may be written

$$\left(\frac{\partial H}{\partial x}\right)_{x\to s^+}\left(\frac{ds}{dt}\right) = \frac{-\sigma_t E^2}{\rho} . \tag{3.19}$$

Apart from its possible rôle in a complete computational study, this approximation can be used to estimate analytically the initial development of the core. It is of interest to note that for $s' > s$, $\frac{ds}{dt}$ is essentially positive.

At $x = s'$ the situation is more complicated. If $s = s'$ then (3.18) certainly holds if it gives decreasing s', and it appears that it always holds in the present case. This need not be true in problems where the heat source can become larger in the cooler region, when a sharp boundary could become diffuse. However if $s \neq s'$, the interface may be moving in either direction. For outwards movement, H is continuous:

$$\left(\frac{\partial T}{\partial x}\right)_{x\to s^+} = 0$$

and by differentiation,

$$\rho c\left(\frac{\partial^2 T}{\partial x^2}\right)_{x\to s'^+}\left(\frac{ds'}{dt}\right) = -K\left(\frac{\partial^3 T}{\partial x^3}\right)_{x\to s'^+} .$$

For inwards movement, the situation closely resembles that for outward movement of s:

$$H_- = H_{x\to s'-} > H_0$$

$$\rho(H_- - H_0)\,\frac{ds'}{dt} = K\left(\frac{\partial T}{\partial x}\right)_{x\to s'+}$$

and
$$\frac{dH_-}{dt} = \frac{\sigma_- E^2}{\rho} + \frac{ds'}{dt}\left(\frac{\partial H}{\partial x}\right)_{x\to s'-} .$$

This provides a complete system of equations.

In starting from cold, as V reaches the switching voltage, s' and then s will become non-zero. If V is then held constant, s and s' will become equal and ultimately reach the steady solution found above. The axi-symmetric version of this problem provides a good approximation to an important practical configuration. However a full solution has not been attempted to date. Apart from the complexities of an elegant computer program, and the uncertain accuracy of a smoothed out approach, the diverse time scales involved appear likely to lead to additional difficulty. The main purpose of the program would be to study periodic operation. In general there are four important time scales - in ascending order:

(*a*) switching on (given essentially by (3.19));

(*b*) switching off (the time scale for conduction across the core);

(*c*) imposed period, (which must clearly be greater than (*b*));

(*d*) time to develop steady cycling (the time scale for conduction across the whole switch).

Understanding of the steady state properties and the general nature of cyclic operation fortunately goes a long way towards satisfying the requirements of theory in thermal switching. In an attempt to provide an adequate model for periodic working without tackling the full equations, the following approach has been used. An empirical core radius has been taken, with a parabolic temperature distribution inside this and a heat transfer boundary condition approximating outward transfer. With some tuning of the radius and boundary condition, this can be made to provide a reasonable model of the actual behaviour. This leaves little practical justification for a full calculation.

ACKNOWLEDGEMENT

This paper is published with the permission of the Directors of Pilkington Brothers Limited and Dr. D.S. Oliver, Director of Group Research and Development.

REFERENCES

1. Atthey, D. R., these proceedings.
2. Berglund, C., *I.E.E.E. Trans. Elec. Dev.*, **ED-16**, 432-437 (1969).
3. Boer, K., *Phys. Stat. Sol.*, **4**, 571-595 (1971).
4. Czerny, M. and Genzel, L., *Glastech. Ber.*, **30**, 1-7 (1957).
5. Kröger, C. and Eligehausen, H., *Glastech. Ber.*, **32**, 362-373 (1959).
6. Patankar, S. and Spalding, D., "Heat and Mass Transfer in Boundary Layers," 2nd edition, Intertext (1970).
7. Pugh, A., *Glasteknish Tidskrift*, **23**, 95-104 (1968).
8. Pye, L., Stevens, H. and La Course, W., "Introduction to Glass Science," Plenum (1972).

GEOPHYSICAL PROBLEMS WITH MOVING PHASE CHANGE BOUNDARIES AND HEAT FLOW

Donald L. Turcotte
(Cornell University, Ithaca, New York)

1. INTRODUCTION

Problems involving moving phase change boundaries with heat transfer and diffusion arise in all branches of geophysics. The freezing of bodies of water is one example. This problem motivated the work of Stefan [3] and the classic one-dimensional freezing of a body is known as the Stefan problem.

Phase change boundaries play an important rôle in meteorology. The lower boundary of a cloud represents a case of convection through a phase boundary. In many cases convection is the dominant heat transport mechanism and the temperature in the lower atmosphere lies on an adiabat. The intersection of the adiabat with the Clapeyron curve for the atmosphere defines the base of the cloud. Within the cloud the temperature must lie on the Clapeyron curve and this is known as the wet adiabat.

More complete studies of cloud physics require estimates of turbulent transport and are semi-quantitative. Lack of understanding of nucleation processes and drop formation also hinders quantitative analysis.

The remainder of this paper will primarily concern several problems in solid earth geophysics. First we discuss the freezing of a magmatic intrusion.

2. FREEZING OF AN INTRUSION

The simplified model for a freezing intrusion is to assume that it has the form of an infinite sill. Assuming the sill freezes symmetrically the Stefan solution may be applied both above and below the sill until freezing is completed. The subsequent relaxation of the thermal anomaly is obtained using Laplace's solution to the heat conduction equation. Modifications of this approach include freezing only on the lower boundary of the intrusion. Studies of fossil intrusions indicate that the freezing magma sinks through the intrusion so that layers of magma solidify from bottom to top. In this case the Stefan solution can still be used below the intrusion and above the intrusion a fixed boundary heat transfer problem is solved. Also various intervals between the liquidus and solidus have been considered. These solutions are summarised by Jaeger [2].

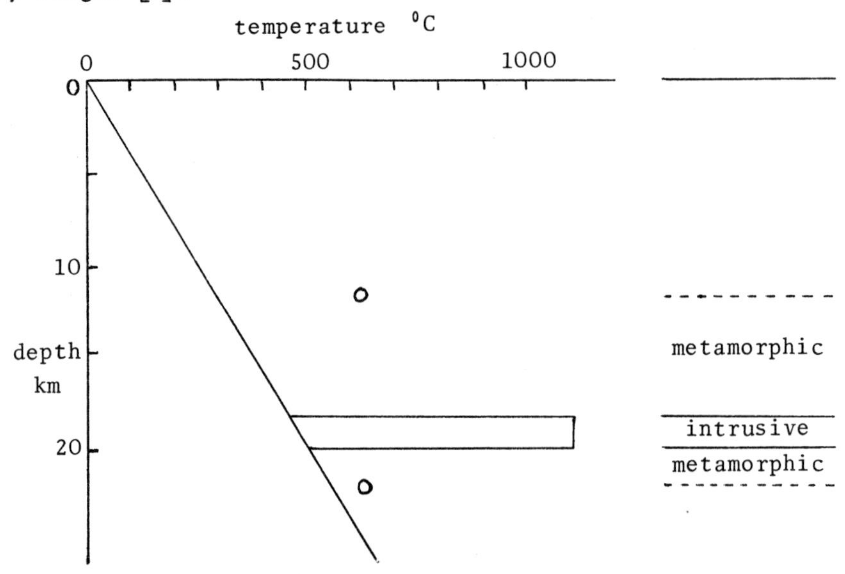

Fig. 1. The solid line gives the temperature as a function of depth immediately after implacement of an intrusion at a depth of 20km, the circles give the extent of metamorphism above and below the intrusion and the inferred maximum temperature at these points.

Observations of metamorphism adjacent to fossil intrusions indicate that the simple conductive heat transfer solutions are not valid. An example is shown in Fig. 1. The circles in this figure show the maximum temperatures experienced at these distances above and below the sill during the cooling of the intrusion. The extent of metamorphism shows a strong preference for upward transport of heat. This can only be explained by convective transport. The extent of the upward transport is so great that the transport must be due to circulation of ground (meteoric) water rather than emission of water or other liquids or gases from the intrusion.

We will return to this problem in the last section of this paper concerning geothermal power. First we discuss some instabilities generated by phase changes in the solid earth.

3. PHASE-CHANGE DRIVEN CONVECTION

Ordinarily a light fluid over a heavy fluid is a stable configuration. However, if the fluids undergo a univariant phase change satisfying a Clapeyron relation and if the system is heated from below the phase change may enhance the Rayleigh instability. This is illustrated in Fig. 2. In Fig. 2(b) a normal Clapeyron relation for a univariant phase change is illustrated. In thermodynamic equilibrium the pressure P_{AB} at which the phase change occurs is related to the temperature by

$$\gamma = \frac{dP_{AB}}{dT} = \frac{L_{AB}}{T(1/\rho_A - 1/\rho_B)} \qquad (3.1)$$

where L_{AB} is the heat evolved per unit mass when the light phase A transforms into the heavy phase B. We consider the situation when γ is positive.

Assume that the two phases are liquids with equal kinematic viscosities ν, equal thermal diffusivities κ, and equal coefficients of thermal expansion α. Consider a horizontal layer of the fluids with a thickness $2d$; in the absence of flow the upper half of the layer is composed of phase A and the lower half is phase B. The layer of fluids is heated from below so that the temperature gradient is β. Two instability mechanisms may generate thermal convection, the ordinary Rayleigh instability associated with the thermal expansion of the fluids and a phase-change instability driven by the density difference between the phases.

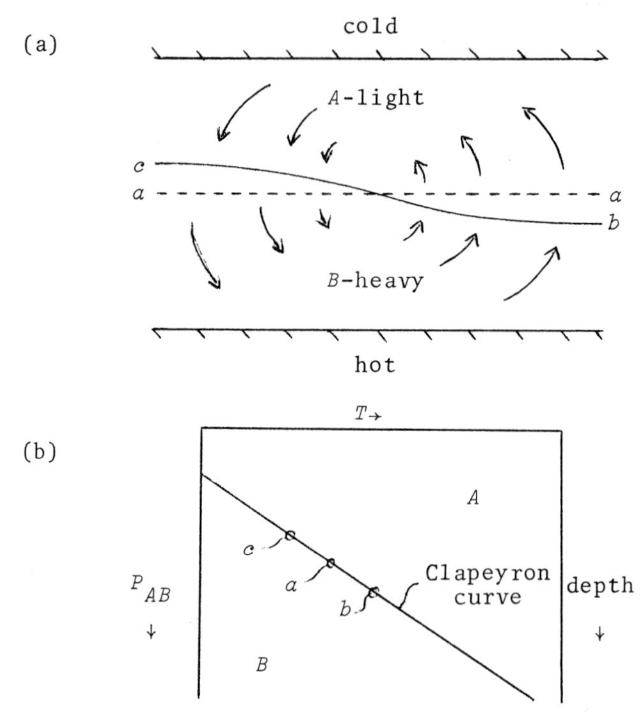

Fig. 2. A fluid layer heated from below initially divided so that the light phase is in the upper half and the heavy phase in the lower half; the displacement of the phase boundary due to a cellular flow is illustrated in (a) and the corresponding points on the Clapeyron curve are given in (b).

The phase-change instability is illustrated in Fig. 2(a). In the absence of convection the phase-change boundary lies at the depth (*a*) at which the hydrostatic pressure and the temperature correspond to the point (*a*) on the Clapeyron curve, Fig. 2(b). An upward flow in the fluid layer brings hotter material towards the phase change boundary as shown on the right side of Fig. 2(a). However, assuming thermodynamic equilibrium the higher temperature drives the phase-change boundary to a higher pressure (*b*) along the Clapeyron curve, and since the pressure is primarily hydrostatic the phase boundary is driven to a greater depth (*b*) as shown. Similarly a downward flow on the left side of the convection cell brings in cold fluid which drives the phase boundary to lower pressure (*c*) and therefore it moves upward to a shallower depth (*c*). Because the elevation of heavier fluid in the region of downward flow tends to hydrostatically drive the flow downwards and the depression of the heavier fluid in the region of upward flow tends to hydrostatically drive the flow upwards an instability results.

However the phase change from B to A in the upward flow absorbs heat and the phase change A to B in the downward flow releases heat. The latent heat associated with the phase change inhibits the instability. In order to determine which effect dominates a detailed analysis is required.

The problem posed above has been solved by Schubert and Turcotte [4] for free surface velocity boundary conditions. The results of the linear stability analysis can be expressed in terms of three dimensionless parameters:

$$R_\beta = \frac{\alpha\beta g d^4}{\kappa\nu} \tag{3.2}$$

$$R_Q = \frac{\alpha L_{AB} g d^3}{c_p \kappa\nu} \tag{3.3}$$

$$S = \frac{\Delta\rho}{\rho\alpha d}\left(\frac{\rho g}{\gamma} - \beta\right)^{-1} . \tag{3.4}$$

The parameter R_β is the usual Rayleigh number based on the coefficient of thermal expansion α. The parameter S is a measure of the importance of the phase change instability to the Rayleigh instability. The parameter R_Q is a measure of the stabilising influence of the latent heat associated with the phase change. The results of the stability analysis are given in Fig. 3.

For small values of R_Q the release of latent heat is unimportant and the stability of the fluid layer is determined by R_β and S. Increasing values of S decrease the critical Rayleigh number for the onset of thermal convection. For large values of R_Q the influence of latent heat in stabilising the flow dominates over the influence of S. For a given value of S there is a value of R_Q which inhibits the phase change instability.

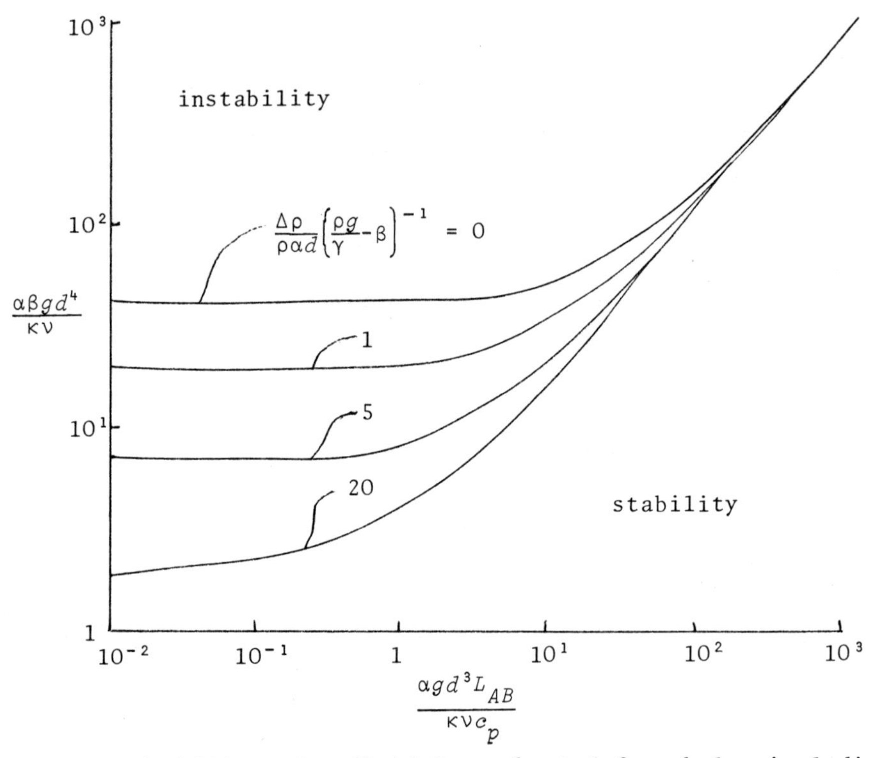

Fig. 3. Stability of a fluid layer heated from below including the effects of a phase change.

An application of this instability is to thermal convection in the earth's mantle. The rigid outer shell of the earth is broken into a number of plates which are in relative motion with respect to each other. These plates are in fact the cold thermal boundary layers of thermal convection cells in the mantle [6]. On long time scales the mantle rocks behave like Newtonian fluids because of solid-state creep. The cold, dense thermal boundary layer is gravitationally unstable and descends into the mantle at ocean trenches. This is illustrated in Fig. 4. It is known both from seismic observations and from laboratory studies that olivine (Mg_2SiO_4, Fe_2SiO_4), the principal component of mantle rocks, undergoes a phase change to a spinel structure at a depth of about 400km. Because of the effect discussed above the phase change is elevated in the cold descending plate as shown in Fig. 4.

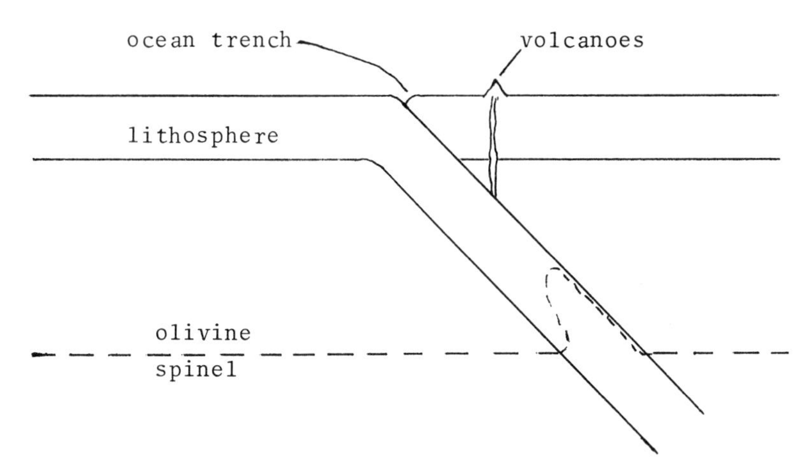

Fig. 4. Illustration of the elevation of the olivine-spinel phase boundary in the cold descending plate.

Since the spinel is about 10% denser than the olivine a body force is exerted on the descending plate. This is the instability mechanism described above. Calculations ([4] and [5]) show that the phase change is elevated about 150km and the resultant body force is of the same order as the gravitational body force due to thermal contraction. This instability mechanism is in part responsible for driving continental drift.

Another phase change that can generate this type of instability in the mantle is melting. Beneath ocean ridges there is ascending convection; due to the decrease in pressure the melting temperature is reduced and melting occurs. This process is identical to condensation in the atmosphere. Since the molten rock is less dense there is an upward buoyancy force which enhances the instability. As in the case described above the melting phase change absorbs heat which tends to stabilise the flow. Again the relative importance of the density change and the heat release must be considered. In the case of a solid-liquid phase change segregation of the liquid may occur. When sufficient quantities of liquid are formed they form liquid inclusions. When the inclusion is sufficiently large the gravitational buoyancy will be sufficient to cause the liquid to rise through the deformable rock as an air bubble rises in water. This process is believed to be important in delivering magma to volcanoes.

4. FLOW THROUGH A PHASE CHANGE

A related problem of geophysical interest is that of flow through a univariant phase boundary. Consider the one-dimensional vertical flow through a phase boundary whose location is determined by the temperature of the material and the hydrostatic pressure. It is assumed that the phase change is in thermodynamic equilibrium and the applicable Clapeyron curve is shown in Fig. 5. For very slow flows the problem is isothermal, essentially identical to the static state without flow. It is assumed that there is no heat conduction at large distances from the phase-change boundary. The phase boundary is a discontinuity at the intersection of the isotherm and the Clapeyron curve. This is case (1) in Fig. 5.

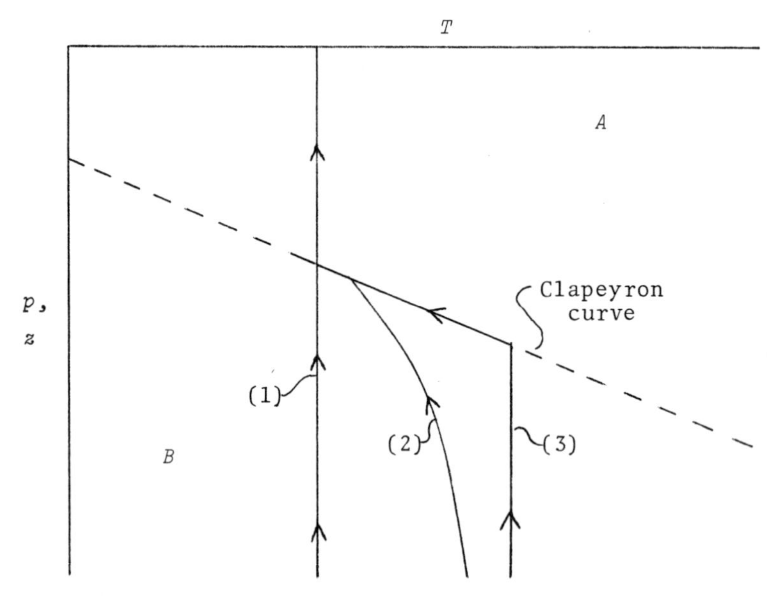

Fig. 5. Upward flow through a univariant phase change; (1) slow flow (isothermal) limit, (2) transition case with a phase discontinuity, (3) fast flow (adiabatic) limit with a two-phase region.

For upward flow the phase change is endothermic and the temperature falls as the phase change occurs. For rapid flow through the phase boundary the process will be adiabatic; there will be no transport of heat by conduction. There will be a two-phase region with a temperature gradient corresponding to the slope of the Clapeyron curve. This is case (3) in Fig. 5 and is completely analogous to the wet adiabat in a cloud as discussed above.

It has been shown by Turcotte and Schubert [7] that the transition from a discontinuity between the phases to a two phase region occurs at a critical velocity V_c given by

$$V_c = \frac{k_A g}{\gamma T_A (s_A - s_B)} \tag{4.1}$$

where k_A is the thermal conductivity of phase A and $s_A - s_B$ is the entropy change associated with the phase change. This transition flow is case (2) in Fig. 5. For all finite velocities the flow is isothermal downstream of the phase boundary. At flow velocities less than the critical velocity a phase discontinuity exists; for flow velocities greater than the critical value a two-phase region exists. The various cases are illustrated in Fig. 6.

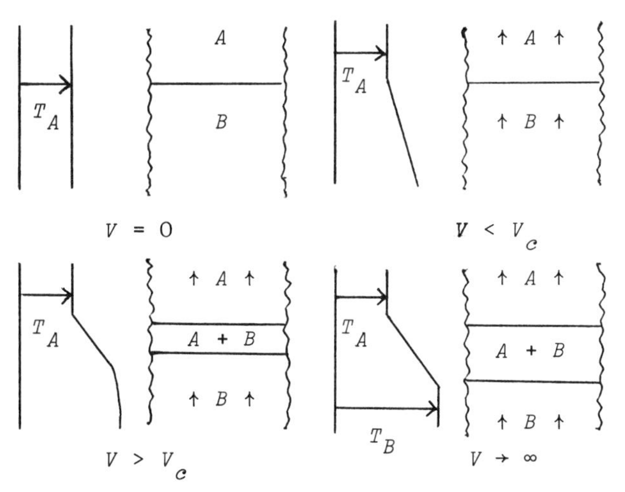

Fig. 6. Temperature profile and phase discontinuity for $V = 0$; conduction upstream and isothermal flow downstream with a phase discontinuity for $V < V_c$; conduction upstream and isothermal flow downstream with a two-phase zone for $V > V_c$; isothermal flow upstream and downstream with two-phase zone for

The critical velocity for the olivine-spinel phase change is 0.2cm/year. This compares with a maximum velocity associated with mantle convection of about 8cm/year.

5. GEOTHERMAL POWER

In the Geysers area north of San Francisco shallow wells strike areas of dry steam. The same is true in the Larderello area in Italy. In a number of other geothermal areas wells strike saturated steam or hot water that can be flashed to steam in the wells. Dry steam is preferable for power generation since the concentrations of corrosive salts and acids are low. In the Geysers the current production of geothermal power is 400Mw with an estimated capacity of 4000Mw. The current production at the Larderello field is 365Mw.

Production of geothermal power from saturated steam depends largely on the corrosive potential of the steam. The Wairakei field in New Zealand yields relatively clean steam and 160Mw of power are produced. It is estimated that the potential power production from the Imperial Valley of Southern California is 20 000Mw. However, the corrosiveness of the wet steam is so great that it has not been possible to extract power for commercial use.

At the present time the understanding of a geothermal field is poor. It is generally agreed that the heat comes from a cooling magmatic intrusive. The steam results from the heating of ground (meteoric) water. Dry steam is found where the porosity of the rocks is low so that the circulation of ground water is small and the water can be heated above its boiling point. The current state of knowledge has been summarised by White *et al* [8].

The depth of the circulation is not known but field studies such as that illustrated in Fig. 1 indicate that the circulation may reach the intrusion even when it is as deep as 20km. There is no direct evidence for the presence of an intrusion beneath the major geothermal fields, so the depth of the magma (assuming there is magma) is not known.

There have been a number of studies of thermal convection in porous media [1]. However, these have not included a phase change. A schematic diagram of the two phase circulation above an intrusion and the resultant geothermal flow is shown in Fig. 7. The circulating water is heated above the intrusion and changes phase. Because of the low density of the steam the hydrostatic head above the intrusion is low. As a result water is driven into this region by the large hydrostatic head where there is no steam. This is exactly the instability illustrated in Fig. 2.

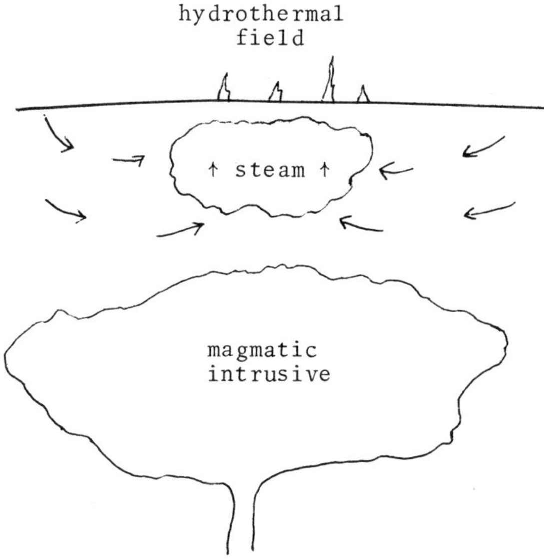

Fig. 7. Schematic diagram of two-phase flow associated with a hydrothermal field.

Many questions remain to be answered. How does the porosity of the rock vary with depth? What happens when the water is above its critical point? Can the production of a geothermal field be altered by injection or removal of water? Estimates of the potential of a geothermal field are imprecise and techniques for finding areas with steam are not well developed.

REFERENCES

1. Elder, J.W., "Physical processes in geothermal areas," *Am. Geophys. Un. Mon.* **8**, 211-239 (1965).

2. Jaeger, J.C., "Thermal effects of intrusions," *Rev. Geophys.* **2**, 443-466 (1964).

3. Stefan, J., "Über die Theorie der Eisbildung, insbesondere über die Eisbildung im Polarmeere," *Ann. Phys. u. Chem.* **42**, 269-286 (1891).

4. Schubert, G. and Turcotte, D.L., "Phase changes and mantle convection," *J. Geophys. Res.* **76**, 1424-1432 (1971).

5. Schubert, G., Turcotte, D.L. and Oxburgh, E.R., "Phase change instability in the mantle," *Science* **169**, 1075-1077 (1970).

6. Turcotte, D.L. and Oxburgh, E.R., "Mantle convection and the new global tectonics," *An. Rev. Fluid Mech.* **4**, 33-68 (1972).

7. Turcotte, D.L. and Schubert, G., "Structure of the olivine-spinel phase boundary in the descending lithosphere," *J. Geophys. Res.* **76**, 7980-7987 (1971).

8. White, D.E., Muffler, L.J.P. and Truesdell, A.H., "Vapor-dominated hydrothermal systems compared with hot-water systems," *Econ. Geol.* **66**, 75-97 (1971).

Astrophysical Problems: A Moving Boundary Problem in the Study of Stellar Interiors

Peter P. Eggleton'
(Institute of Astronomy, Cambridge)

Over the last 50 years a physical picture has been developed of the interiors of stars, and of their evolution with time. This picture is quite detailed, and appears to be able to explain reasonably well the (rather few) directly measurable properties of the majority of stars. Much of this detailed picture has come in the last fifteen years, with the aid of large computers [5], [6], [7] and [8].

The equations which we believe are adequate are few and simple, at least if we restrict ourselves to spherically symmetric (*i.e.* non-rotating, non-magnetic, isolated) stars. Their numerical solution however poses some problems, even if one has a very generous allocation of computer time. To start with, we have, in a Lagrangian formulation [9]:

mass conservation: $$\frac{\partial r}{\partial m} = \frac{1}{4\pi r^2 \rho}, \tag{1}$$

hydrostatic equilibrium: $$\frac{\partial p}{\partial m} = -\frac{Gm}{4\pi r^4}, \tag{2}$$

heat flux: $$\frac{\partial T}{\partial m} = -\frac{\alpha L}{4\pi r^4}, \tag{3}$$

energy conservation: $$\frac{\partial L}{\partial m} = \varepsilon_{nuc} - T\frac{\partial s}{\partial t}, \tag{4}$$

fuel consumption: $$\frac{\partial X}{\partial t} = -\frac{\varepsilon_{nuc}}{E}, \tag{5}$$

where r is the radial coordinate, m the mass up to that radius, ρ is the density, G the gravity constant, p the pressure, T the temperature, S the entropy, L the luminosity (*i.e.* total heat flux), ε_{nuc} the nuclear energy generation rate, X the concentration of fuel (usually hydrogen, which burns to helium), and E the energy content of the fuel. The quantities ρ, s and ε_{nuc} are themselves functions of the thermodynamic state of the gas, as defined by, say, p, T and X. The equations (1) to (5) are five equations in the five "independent" variables r, p, T, L and X.

The transport coefficient α is the only problem at this stage. Normally, it is simply the radiative opacity of the material, which is also a function only of the thermodynamic state of the gas. But this opacity may be so large in some regions that it demands a temperature gradient considerably steeper than the adiabatic gradient. In this situation we expect turbulent convective motion to be set up. Such convection is usually so efficient at carrying heat that it reduces the temperature gradient to a value fractionally in excess of the adiabatic value. This can therefore be accommodated (at least crudely) in the equations by modifying α as follows [1].

Let $$\alpha_r \equiv f(p, T, X) \equiv \text{radiative opacity}, \tag{6}$$

$$\alpha_a \equiv \frac{Gm}{L}\left(\frac{\partial T}{\partial p}\right)_{s,X} \equiv \text{convective "opacity"}. \tag{7}$$

Then the actual transport coefficient α is given by

$$\begin{aligned} \alpha &= \alpha_r && \text{if } \alpha_r \leqslant \alpha_a , \\ &= \alpha_a + \delta(\alpha_r - \alpha_a) && \text{if } \alpha_r \geqslant \alpha_a , \end{aligned} \tag{8}$$

where δ is very small (usually), though not accurately known.

This convective motion affects not only the heat transport. It mixes the material within the unstable region sufficiently rapidly that it prevents a composition gradient from building up there, even though the nuclear reactions are working to produce a rather steep gradient. We have to modify the composition equation to something like this [3]:

$$\frac{\partial}{\partial m}\left(\sigma \frac{\partial X}{\partial m}\right) = \frac{\partial X}{\partial t} + \frac{\varepsilon_{nuc}}{E} , \tag{9}$$

with $$\begin{aligned} \sigma &= 0 && \text{if } \alpha_r \leqslant \alpha_a , \\ &= K(\alpha_r - \alpha_a) && \text{if } \alpha_r \geqslant \alpha_a , \end{aligned} \tag{10}$$

where K is very large.

The structure equations and the composition equation can be written together as a set of six first order differential equations (drawing a veil over the discontinuous character of some of the functions involved) of the following character:

$$\frac{\partial}{\partial m} g_i(y_j) = f_i\left(y_j, \frac{\partial y_j}{\partial t}\right) . \tag{11}$$

These can be approximated by an implicit difference scheme of the form:

$$\begin{aligned}\frac{g_i(y_j^{(k)}) - g_i(y_j^{(k-1)})}{\Delta m^{(k-\frac{1}{2})}} &= \alpha_i\, f_i\left(y_j^{(k)}, \frac{y_j^{(k)} - Y_j^{(k)}}{\delta t}\right) \\ &+ (1 - \alpha_i)\, f_i\left(y_j^{(k-1)}, \frac{y_j^{(k-1)} - Y_j^{(k-1)}}{\delta t}\right) ,\end{aligned} \tag{12}$$

where $Y_j^{(k)}$ is the value of $y_j^{(k)}$ in the previous time step. The α_i are weights which have to be chosen rather carefully to obtain both accuracy and numerical stability. This set of difference equations, along with some straightforward boundary conditions at the star's centre and surface, can be differentiated numerically (the functions involved being often very complicated) and solved for the $y_j^{(k)}$ by Newton-Raphson. Given a starting model, say a star of uniform composition with 70% hydrogen, 30% helium and a trace of all other elements, we can follow its development in time.

Fig. 1 shows the thermodynamic state of the gas in stellar interiors, as it emerges from such computations [4]. Most model stars are found to occupy a region

$$10^{-15} \lesssim \rho/T^3 \lesssim 10^{-7} ,$$

$$\text{and} \qquad 3.10^3 \lesssim T \lesssim 10^9 .$$

Individual stars tend to lie on a locus of roughly $\rho \propto T^3$, with more luminous stars having a higher temperature at a given density. Most stars also tend to move upwards in this diagram as they evolve. It is a fortunate circumstance that most stars exist for most of their life within the dash-dot boundary, which marks out various phase changes that can occur in different temperature-density regimes. If a star strays into one of the adjacent regions it can be expected either to collapse (to a black hole), explode (as a supernova), or solidify and cool off (to a white dwarf or neutron star).

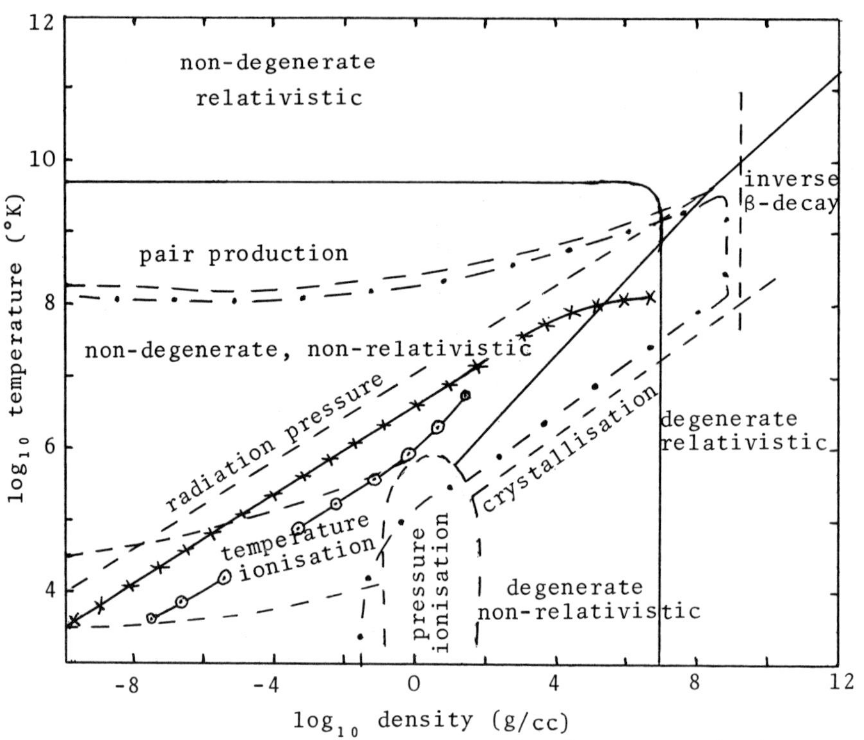

Fig. 1. The equation of state in the (logρ, log$T$) plane, assuming a hydrogen/helium mixture. Solid lines (———) separate regions in which the electron gas has various limiting types of behaviour. Dashed lines (-----) indicate boundaries at which various physical effects become important. The dash-dot line (-·-·-) defines the region in which the material behaves nearly as a perfect gas. The position occupied by material at different points in the sun is indicated thus (⊙—⊙—⊙), and, in a more luminous star, thus (✕—✕—✕).

However it is not the collapse or solidification of stars at the end of their active lives that we wish to discuss, but rather the problem of locating those moving boundaries within a star which separate the convectively unstable regions from the stable regions. Fig. 2 illustrates the evolution of a typical star, with time plotted horizontally (or, more

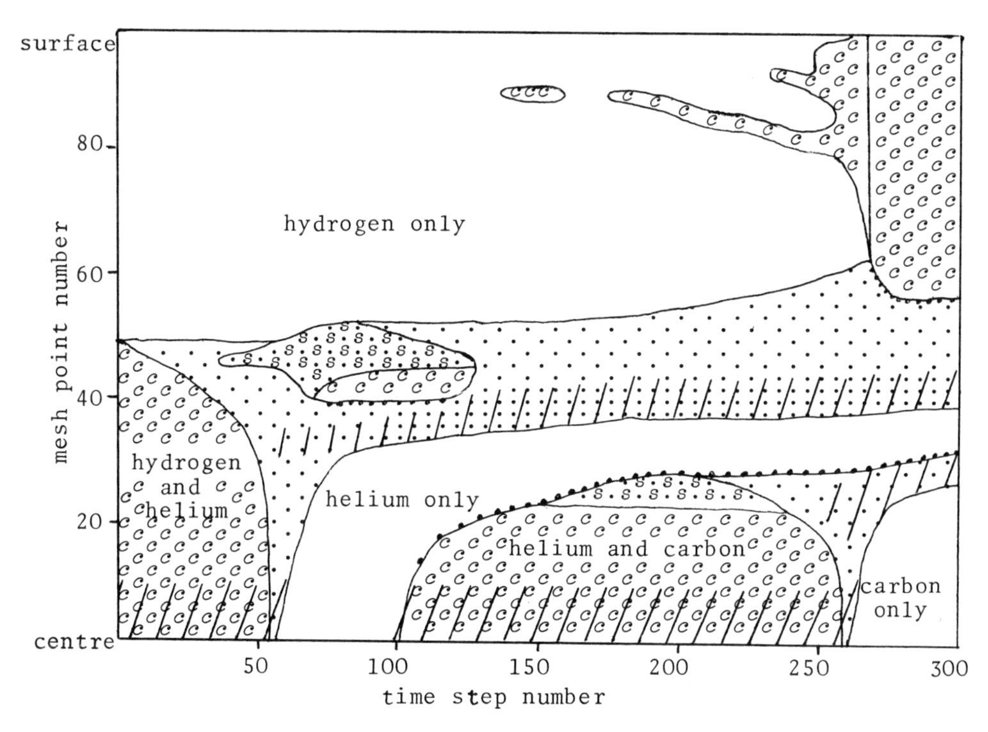

Fig. 2. The evolution of a typical star, from the main sequence to the formation of a carbon core. Time steps are plotted horizontally, space steps vertically. The time steps are not equal; the mass steps are not equal and are not the same at different times. Convective and semi-convective regions are denoted by *c* and *s*. Density of dots indicates roughly the steepness of the composition gradient. Diagonally striped regions indicate where nuclear burning principally occurs.

---

precisely time step number; the time steps are necessarily of very unequal lengths), and radial coordinate plotted vertically (more precisely, spatial mesh point number; the spatial intervals also vary widely in size, and furthermore vary with time). Convective regions, which appear and disappear during evolution, are marked with the letter *c*. Stable regions are blank, or else covered with dots in regions where there is a composition gradient. But it is interesting to find that computer solutions of the composition diffusion equation (9), with diffusion coefficient as given by (10), from time to time demand the existence of a third kind of region, marked with *s* for semi-convective. Normally, the fact that the coefficient K in equation (10) is very large means that the composition gradient $\partial X/\partial m$ is very small (in a region where $\alpha_r > \alpha_a$). But in a semi-convective region the fact that K is large implies that it is $\alpha_r - \alpha_a$ which is very small, while the composition gradient is of order unity. It is very satisfactory that the implicit difference scheme based on equation (9) along with equations (1) to (4) introduces such regions naturally without having to be told to do so.*

A further complication in the numerical treatment of stars (as of many other systems) is the choice of distribution of mesh points. There is no way of dividing up a star into fixed, even if unequal, intervals of mass coordinate which will be uniformly satisfactory at all stages of evolution. For instance, the pressure may change by one per cent across a given portion of matter at one stage, and by a factor of 10^6 across the same portion at a later stage. One is therefore driven to a non-Lagrangian mesh, where a given portion of matter can move backwards or forwards through the mesh, though fixed in the star. A procedure which is very simple to apply, because it involves simply throwing two extra difference equations of the same character as the others (equation (12)) into a "black box" which solves any given number of similar simultaneous difference equations, is the following. Think of some function q, say, definable (except for an arbitrary multiplicative factor) at each mesh point as a function of the independent variables which in the stellar case are X, r, p, T, L and the composition gradient. This function must have the following properties:

(*i*) it is monotonic from the centre to surface and increases by unity over the whole range (this latter condition can be achieved using the arbitrary multiplicative factor, which does not need to be known in advance, and which may vary with time);

(*ii*) in a region where q varies by, say, one per cent no other quantity of physical interest changes by more than a few per cent.

---

*See also Atthey (p.187).

As an illustration, the quantity

$$q = C(\log m - \log p)$$

has roughly this property in a stellar problem (provided we make the central boundary at some small distance away from $m = 0$). Practice shows that this "coordinate" divides up the star into reasonable zones. Then we have two extra differential equations to solve [2],

$$\frac{\partial q}{\partial m} = C\left\{\frac{1}{m} + \frac{Gm}{4\pi r^4 p}\right\} > 0, \tag{13}$$

and

$$\frac{\partial C}{\partial m} = 0. \tag{14}$$

The last equation may appear rather trivial, but it needs to be of the same form as the others in order to go in the standard "black box". The equations can now be recast as eight simultaneous equations in the eight independent variables r, p, T, L, X, m, C and composition gradient, of the form:

$$\frac{\partial}{\partial q} g_i(y_j) = f_i\left(y_j, \frac{\partial y_j}{\partial t} - \frac{\partial y_j}{\partial m}\frac{\partial m}{\partial t}\right), \tag{15}$$

and discretised over a mesh at equal intervals of q in the range $\{0,1\}$. Time derivatives are now at constant mesh point. This recipe appears to be able to re-distribute mesh points automatically to where they are most useful. However, in a situation where a thin front of some sort develops and begins to move, perhaps rapidly, through the star, some care must be taken in choosing which of several superficially similar difference approximations to (15) is used.

Within this framework, what should one do about the moving boundaries between convective, semi-convective and stable regions? In our experience, nothing. Provided common sense is exercised, it seems adequate to act as if the functions α, in (8) and σ, in (10), are quite smooth in the neighbourhood of $\alpha_r = \alpha_a$. The fact that their derivatives have jump discontinuities, and that σ may vary by several orders of magnitude across a single interval, does not affect the convergence, or the reliability very much. Indeed the physical model underlying the composition diffusion equation (9) is really only plausible when applied on a macroscopic length scale, and cannot be justified on the microscopic scale implicit in the use of a differential equation. Thus we may say that it is actually the difference equation that models the composition mixing better than the differential equation, and solutions of the difference equation appear to be quite well behaved. Fig. 3 illustrates the sort of composition profile that may arise in a numerical treatment of a convective and semi-convective region.

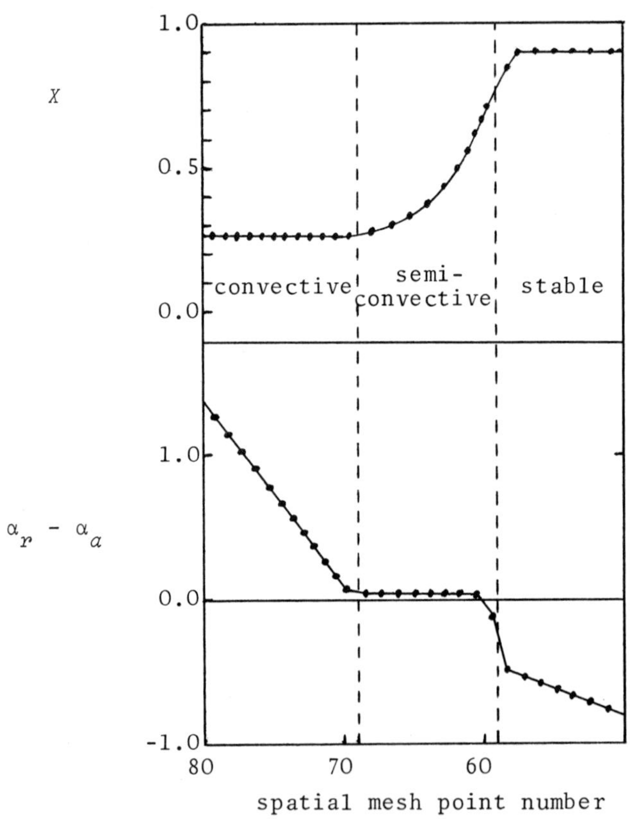

Fig. 3. The composition X and the convective stability criterion $\alpha_r - \alpha_a$ through part of a model star. The position of the boundaries (- - -) can move with time, and are moderately well defined.

Emphasis has been placed on the good points of this kind of treatment, but there is the difficulty that one can never be sure how difference equations of this order of complexity will work in practice, and can hardly hope to quantify the sort of accuracy that is obtained. The most fundamental check is simply that the results should accord with simple order-of-magnitude estimates that one can make on purely physical grounds.

REFERENCES

1. Baker, N.H. and Ternesvary, S., "Tables of Convective Stellar Envelopes," Goddard Institute for Space Studies (1966).
2. Eggleton, P.P., *Mon. Not. Roy. Astr. Soc.* **151**, 351 (1971).
3. Eggleton, P.P., *Mon. Not. Roy. Astr. Soc.* **156**, 361 (1972).
4. Eggleton, P.P., Faulkner, J. and Flannery, B.P., *Astron. and Astrophys.* **23**, 325 (1973).
5. Härm, R. and Schwarzschild, M., *Astrophys. J.* **145**, 496 (1966).
6. Iben, I., *Astrophys. J.* **143**, 505 (1966).
7. Iben, I., *Astrophys. J.* **147**, 624 (1967).
8. Paczynski, B., *Acta. Astr.* **20**, 47 (1970).
9. Schwarzschild, M., "Structure and Evolution of the Stars," Princeton University Press, (1958).

Contribution Concerning the Solidification Problem

D. J. Hebditch
(C.E.G.B., Marchwood Engineering Laboratories)

In the paper by Perkins (p.19), reference was made to the practical importance of the interaction of heat and mass transfer. The particular case reported was that of the melting of scrap steel in a liquid iron bath. At the later stages of dissolution, the liquidus temperature of the scrap steel was higher than the temperature of the bath, yet dissolution went to completion due to carbon transport by diffusion and convection.

In general, however, little experimental or theoretical work has been reported on coupled heat and mass transfer at moving boundaries. This contribution describes briefly some experimental work [1] and [2] which shows simultaneous heat and mass transfer in the solidification of a binary alloy. In particular the work provides information on the experimental variables as a function of time. The variables are composition, temperature and position of the dendritic boundary.

Casts of height 6 cm, thickness 1.3 cm and length 10 cm were made by extracting heat, at approximately constant rate, horizontally and unidirectionally from one edge of the liquid alloy. The alloy was initially at the liquidus temperature. Considerable care was taken to minimise heat flow through all surfaces except the cooled one. Unidirectionality of heat flow is demonstrated by the macroscopically flat quenched interface of alloy *A*, shown in Fig. 1. Alloy *A* was Pb-Sn eutectic (*i.e.*, temperature invariant freezing point, see Fig. 4), so that the interface represents an isotherm, which is almost planar and parallel to the cooled end surface.

Fig. 1 Macroscopically planar interface in a partially grown Pb-Sn eutectic cast, alloy *A*; horizontal and vertical sections.

Fig. 2 Sn-5 wt.pct.Pb partially grown cast, alloy *B*, growth left to right.

(*a*) Macrostructure.

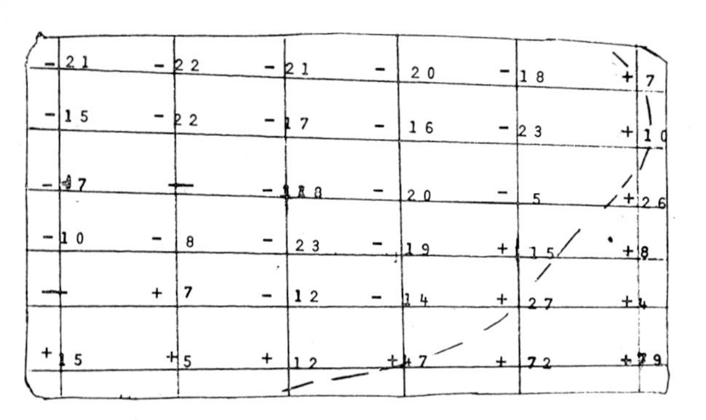

(*b*) Percentage macrosegregation diagram (dendrite interface shown dashed).

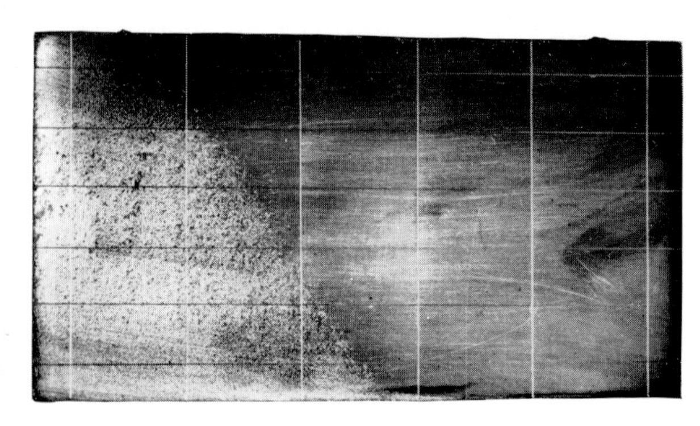

Fig. 3 Pb-48 wt.pct.Sn partially grown cast, alloy *C*, growth left to right.

(*a*) Macrostructure.

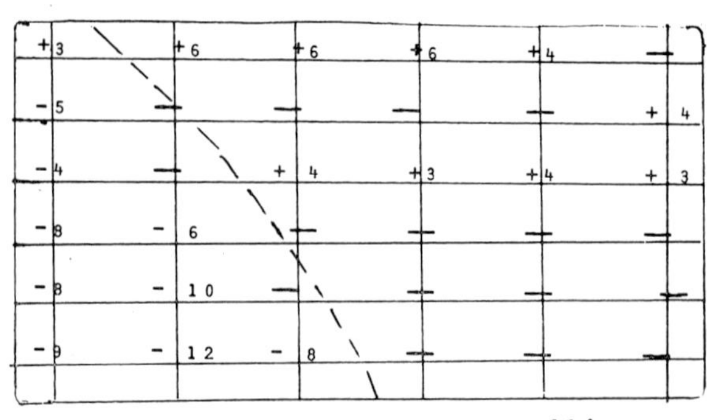

(*b*) Percentage macrosegregation diagram (dendrite interface shown dashed).

In Fig. 2a, a dendritic interface, quenched before the end of solidification, is shown for alloy *B*, Sn-5wt.pct.Pb (see Fig. 4). This composition is far off eutectic and the interface is non-isothermal. In Fig.2b the average analyses at points for the cast of Fig. 2a are shown. The analyses are given as positive or negative percentage deviations from the nominal composition. The alloy was homogeneous before solidification, so that the interface curvature and the inhomogeneity are due to freezing. Figs. 3a and 3b show the macrostructure and compositions for alloy *C*, Pb-48wt. pct.Sn (again see Fig. 4), cast under the same conditions. The slope of the dendritic interface is quite different from that of alloy *B*.

The dominant mechanism causing the interfacial curvature and composition inhomogeneity is slow convective flow for both alloy *B* and alloy *C*. The direction of flow in the dendritic mushy region can be predicted from the liquid density variation with composition and the phase diagram. Thus for alloy *B*, the interdendritic liquid becomes enriched in Pb, and more dense than the bulk liquid ahead of the advancing interface. This results in convective flow, downwards in the dendritic region and upwards in the completely liquid region. For alloy *C*, the interdendritic liquid becomes less dense, and so flow is in the opposite sense to that for alloy *B*. Flow of liquid in the mushy region can occur in equilibrium with the adjacent solid but under other circumstances the flow may be unstable leading to fewer but larger channels in which the liquid may no longer be at equilibrium with the local solid. In either case local variations in fraction solid occur which cause macrosegregation.

It is emphasised that the casts shown in Figs. 2a and 3a are drawn from series of casts of similar composition and thermal conditions but which by use of the quenching technique, represent various times after the start of cooling.

This work provides a systematic experimental investigation of a coupled heat and mass transfer situation. The work could form the basis for the construction of a mathematical model which would have considerable importance in the fields of ingot solidification and casting.

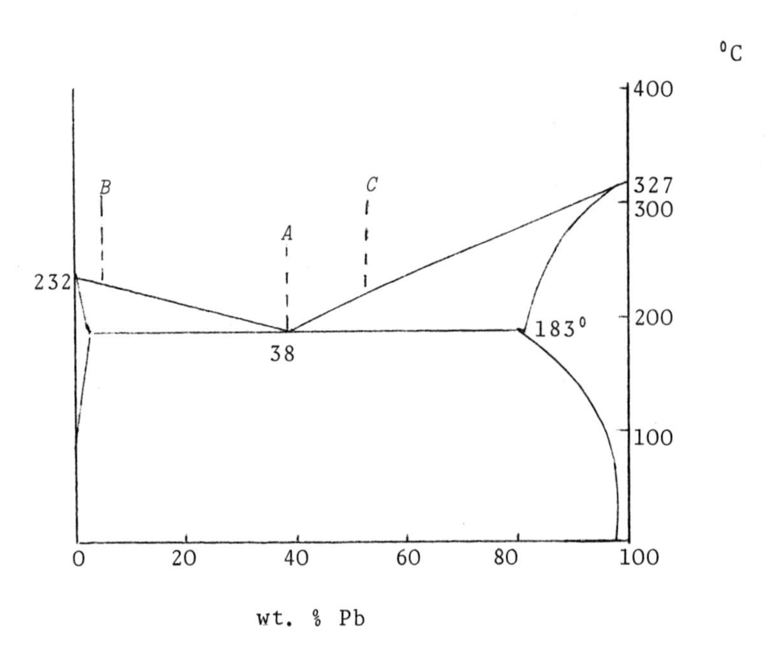

Fig. 4 Pb-Sn phase diagram illustrating compositions of alloys *A*, *B* and *C*, respectively, 38, 5 and 52 wt.pct.Pb.

REFERENCES

1. Hebditch, D.J., D. Phil. Thesis, Oxford University (1973).
2. Hebditch, D.J. and Hunt, J.D., "Observations of Ingot Macrosegregation on Model Systems", in publication, *Met. Trans.* (1974).

PART 11. ANALYTICAL AND NUMERICAL METHODS

INTRODUCTION

The analytical and numerical section of this book is chiefly concerned with the practical solution of moving boundary problems of the type described in Part I. The ideas presented are some of those which a research worker should have in his mind when attempting to obtain a sensible mathematical solution to such a problem.

It is impossible to produce such a solution without beginning with a correct mathematical formulation and in the first paper in this section, Tayler describes the simplest way in which the problem may be posed. This is as a boundary value problem for the heat conduction equation and Tayler also describes several transformations of the variables in this boundary value problem which may simplify the subsequent analysis. As in the rest of this section, attention is directed towards problems dominated by heat transfer and diffusion; effects such as convection are either ignored or modelled very crudely. Even within this area it is pointed out that the boundary value problem formulation may actually preclude the only physically acceptable solutions. These are solutions which involve so-called mushy regions; such regions are discussed later in more detail by Atthey and Langham.

In simple cases, the weak version of the boundary value problem formulation offers a generalisation which is just wide enough for physically acceptable solutions to be obtained. This concept is described by Tayler and in more detail by Atthey but its usefulness in more complicated situations, say involving simultaneous heat and mass transfer, is still unclear.

Once one formulation has been written down which admits physically acceptable solutions, various reformulations are possible which may either make it easier to discover if the problem is well posed mathematically or to derive analytical or numerical results. One well established reformulation of the boundary value problems is in the form of Volterra integral or integro-differential equations for the phase boundary, and this approach is described in the paper by Ockendon. One method by which this reduction may be achieved, involving embedding the moving boundary problem in a problem with fixed boundaries, was devised by Boley in 1960. In his paper here he describes the use of the embedding technique in a wide variety of situations. The papers of both Ockendon and Boley also describe some of the numerous asymptotic estimates which are available for Stefan problems; the limits of large or small latent heats may be considered, as well as local and far-field expansions.

Quite a different kind of reformulation of moving boundary problems uses a variational framework. Although no true variational principle exists for Stefan problems, the temperature field can be shown to satisfy a variational inequality and Duvaut uses this idea to discuss the existence and uniqueness of the solution. A different variational approach, described by Agrawal in the discussion section, uses Biot's principle. Here the variations in a suitably defined thermal potential and dissipation function are related to a generalised virtual work which is the product of the temperature and the normal heat flux at the boundary. The Stefan problem is thus reduced to the integration of the associated Lagrangian equations.

It is one of the interesting features of moving boundary problems that different mathematical formulations suggest different, but practicable numerical schemes. To find weak solutions, Atthey describes a finite difference scheme first devised by Oleinik in which the temperature and heat content are used as simultaneous dependent variables. Duvaut's formulation is naturally implemented by an iterative procedure in which a minimisation is carried out at each time step. Other formulations may be computed in a more conventional manner.

Of the numerous numerical schemes for Stefan problems, finite difference methods are perhaps the most popular and Crank's paper describes their application to a variety of problems. Emphasis is placed on the ways in which preliminary transformations or the use of variable meshes can improve their efficiency. A Stefan problem which has frequently been used to test and compare different numerical schemes is the oxygen diffusion model of Crank and Gupta, described in Crank's paper in Part 1. A new finite difference solution to this problem is given in the discussion session by Ferriss.

Another very popular and practical computing technique is Goodman's heat balance method, which is similar to Polhausen methods for calculating boundary layer flows. This method was discussed at the conference by Noble and a brief resume of his paper is included here.

Finally a survey is given by Fox of these and other numerical methods which have appeared in the literature, along with some guidelines on which techniques are likely to be most valuable in any particular situation.

Since the emphasis throughout this section is on practical methods of solving Stefan and related problems, this list of topics does not cover the mathematical literature of these problems by any means. In particular, little or no reference is made to the considerable body of results on the existence, uniqueness or even stability of solutions of the various possible formulations, nor of the interesting relationship between free boundary problems and the theory of optimal stopping processes.

THE MATHEMATICAL FORMULATION OF STEFAN PROBLEMS

A.B. Tayler
(University of Oxford)

1. INTRODUCTION

The fundamental feature of a Stefan problem is that the parabolic diffusion equation is satisfied in a region or regions whose boundaries are to be determined. Thus for a three-dimensional continuum at rest in suitable non-dimensional variables the appropriate model equation may be written

$$\frac{\partial u}{\partial t} = \operatorname{div}(\beta \text{ grad } u) + a \tag{1.1}$$

where a and β are piecewise differentiable functions of $\underline{x}$, t and u and the diffusion coefficient $\beta \geqslant 0$. In most of the applications only one space dimension is considered and source densities a are neglected so that

$$\frac{\partial u}{\partial t} = \frac{\partial}{\partial x} \left(\beta \frac{\partial u}{\partial x}\right). \tag{1.2}$$

If β is a constant the elementary solution of (1.2) is

$$t > 0,\ u = (4\pi t)^{-\frac{1}{2}} \exp\left(\frac{-x^2}{4\beta t}\right);\ u = 0 \text{ otherwise},$$

and can represent the temperature distribution in an infinite one-dimensional space due to a unit of heat applied at the origin $x = 0$ at time $t = 0$. In an n-dimensional continuum the elementary solution is [1]

$$t > 0,\ u = (4\pi t)^{-n/2} \exp\left(-\frac{|\underline{x}|^2}{4\beta t}\right);\ u = 0 \text{ otherwise}.$$

If $\beta = u^b$ where $b > 0$ the elementary solution of (1.2) is

$$t > 0,\ \left|\frac{x}{x_1}\right| < 1,\ u = \left(\frac{t}{t_0}\right)^{-\frac{1}{b+2}} \left\{1 - \left(\frac{x}{x_1}\right)^2\right\}^{1/b};\ u = 0 \text{ otherwise},$$

$$\text{where } \frac{x_1}{x_0} = \left(\frac{t}{t_0}\right)^{\frac{1}{b+2}},\ t_0 = \frac{b x_0^2}{2(b+2)} \text{ and } x_0 = \frac{\Gamma(1+\frac{1}{b}+\frac{1}{2})}{\sqrt{\pi}\ \Gamma(1+\frac{1}{b})}.$$

This wave-like behaviour in the nonlinear case, which cannot occur when β is a constant, is due to the vanishing of the diffusion coefficient. In the limit $b \to 0$ the correct solution for β constant is obtained. Similar expressions can be obtained for higher dimensions.

In the most common applications u represents either temperature [3] or the concentration of a solute in a solvent [1] and [5]; other examples are the pressure in a fluid diffusing through a porous medium or the vapour density in an undersaturated mixture. In a more novel application u represents the reward function of a game in which x is the "state of affairs" of a gambler faced with an optimal stopping decision [12]. The continuum in a physical application can have two or more phases in which some material properties are markedly different, the simplest example being when u represents temperature and the continuum can have either a liquid or solid phase. This problem was first discussed by Stefan in relation to the melting of ice caps, and in the classical Stefan problem β is piecewise constant. Most of the interest in the solution of the problem lies in the properties of the boundary, or boundary surface, between the phases. When u represents temperature then latent heat is created or absorbed at the phase change temperature, which by suitable scaling can be taken as $u = 0$. This will give rise to discontinuities in the derivatives of u across the phase change boundary on which $u = 0$. The appropriate discontinuities and the governing diffusion equation can be derived from an integral formulation of the principle of conservation of heat. In one dimension this is

$$\int_C \left\{ A(u)\,dx + \beta\frac{\partial u}{\partial x}dt \right\} = 0 \tag{1.3}$$

for any closed contour C in the x,t plane where

$$A(u) = u \text{ for } u > 0; \; A(u) = u - \lambda \text{ for } u \leqslant 0. \tag{1.4}$$

$A(u)$ and λ represent the heat content and latent heat in appropriate non-dimensional form and λ is positive. Thus in any region where $u \neq 0$ we can rewrite (1.3) as

$$\iint\left\{\frac{\partial}{\partial t}A(u) - \frac{\partial}{\partial x}\left(\beta\frac{\partial u}{\partial x}\right)\right\}dx\ dt = 0$$

over an arbitrary area and hence recover (1.2). However, if $u = 0$ on $x = s(t)$, by considering a contour C which is tangent to a section of $x = s(t)$ (see Fig. 1) in the limits $x \to s(t)+$ and $x \to s(t)-$, then across such a boundary

$$\left[A(u)\right]_1^2\dot{s} + \left[\beta\frac{\partial u}{\partial x}\right]_1^2 = 0,$$

where suffices 1 and 2 denote limiting values in regions $u < 0$ (solid) and $u > 0$ (liquid) respectively. We thus obtain the Stefan condition that

$$x = s(t),\ u = 0,\ \left[\beta\frac{\partial u}{\partial x}\right]_1^2 = -\lambda\dot{s}\ . \tag{1.5}$$

In three dimensions with a phase change surface $F(\underline{x},t) = 0$ the Stefan condition generalises to

$$\lambda\frac{\partial F}{\partial t} = \left[\beta\ \mathrm{grad}\ u\right]_1^2 \cdot \mathrm{grad}\ F\ . \tag{1.6}$$

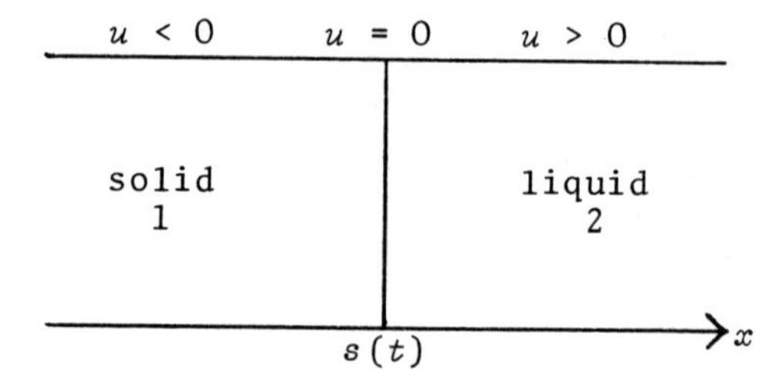

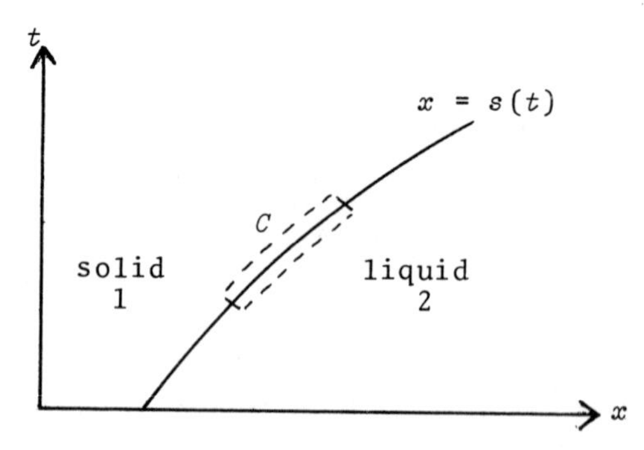

Fig. 1. One dimensional phase change boundary.

This principle of conservation of heat is a simplified version of conservation of energy in which both phases are assumed to be incompressible and at rest. Across any phase change boundary there must be conservation of energy, momentum and mass and the latter then requires that the material density is continuous. For the case of two incompressible phases of different densities ρ_1 and ρ_2, the first of which is a solid, then the liquid phase must have a velocity V to satisfy the mass continuity condition where *

$$\rho_1 \dot{s} = \rho_2(\dot{s} - V) \text{ on } x = s(t). \tag{1.7}$$

In the liquid region it is necessary to add a convective term to the diffusion equation so that

$$\frac{\partial u}{\partial t} + V\frac{\partial u}{\partial x} = \frac{\partial}{\partial x}\left(\beta\frac{\partial u}{\partial x}\right). \tag{1.8}$$

In a one-dimensional problem the continuity and momentum equations for an inviscid liquid require that $\frac{\partial V}{\partial x} = 0$ but in three dimensions the flow problem and consequent convective term $\underline{V} \cdot \text{grad } u$, are considerably more complicated.

2. THE CLASSICAL STEFAN PROBLEM

In the problem of ablation† a solid is melted by a moving fluid in which it can be assumed that the diffusion coefficient is very large so that $\frac{\partial^2 u}{\partial x^2} = 0$ in the liquid. Thus the heat input $q(t)$ from the fluid to the solid is prescribed on the unknown melting boundary $x = s(t)$. This is a one phase Stefan problem and in one space dimension a typical boundary value problem is (see Fig. 2)

$$\left.\begin{aligned}
0 < x < s(t),\ \frac{\partial u}{\partial t} &= \frac{\partial}{\partial x}\left(\beta\frac{\partial u}{\partial x}\right);\\
x = s(t),\ u = g(t),\ \beta\frac{\partial u}{\partial x} &= \lambda\dot{s} + q(t);\\
t = 0,\ u = \phi(x) < 0,\ s &= 1;\\
x = 0,\ u = f(t) &< 0,
\end{aligned}\right\} \tag{2.1}$$

where, in the case of ablation, $g = 0$ and $q > 0$.

The boundary value problem (2.1) is well posed [11] with a unique solution for $t < t_0$, which possesses continuous second

*See, e.g., Gelder and Guy (p.77).

†See Agrawal (p.247).

derivatives (C^2) provided that $f(t)$ is twice differentiable.

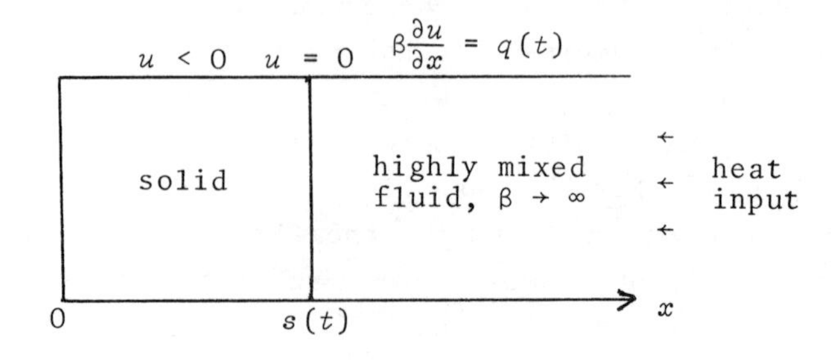

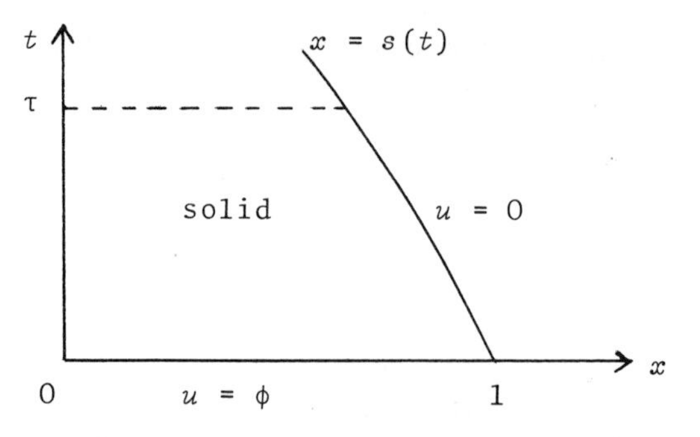

Fig. 2. Ablating solid, one phase.

The restriction implied by t_0 is that the phase may disappear (degenerate) at some finite time and in this sense no solution exists for later times. It is also well posed if $s(0) = 0$ although the proof is more difficult. A necessary condition is that $q(0) < 0$ and is a model of accretion rather than ablation. The proof for $s(0) \neq 0$ consists in reformulating the problem as a pair of Volterra integral equations for $s(t)$ and $v(t) = \lim_{x \to s} \frac{\partial u}{\partial x}$, using the Green and Neumann functions for the diffusion equation on $(0,\infty)$. (The manipulation is similar to that on p.139). They have a unique solution for sufficiently small t and continuity arguments can be used to extend this solution to $t < t_0$. The proof is essentially the same for variable latent heat $\lambda(t) > 0$ and a boundary condition $\mu u - \frac{\partial u}{\partial x} = f(t) < 0$, $\mu > 0$ on $x = 0$. The case $\lambda = 0$ is unexpectedly more difficult than $\lambda \neq 0$ and requires a preliminary reformulation Schatz, [10]. With $q \neq 0$, we put $w = \frac{\partial u}{\partial x}$ so that

$$
\left.
\begin{aligned}
0 < x < s(t), \quad & \frac{\partial w}{\partial t} = \frac{\partial^2 w}{\partial x^2}; \\
x = s(t), \quad w = q(t), \quad & \frac{\partial w}{\partial x} = -q\dot{s} + \dot{g}; \\
x = 0, \quad \frac{\partial w}{\partial x} = \dot{f}(t); \quad & t = 0, \quad w = \phi'(x);
\end{aligned}
\right\} \tag{2.2}
$$

where for simplicity we have taken $\beta = 1$. If $q = 0$ the appropriate transformation is $w = \frac{\partial u}{\partial t}$ provided $\dot{g} \neq 0$. If $\lambda = 0$, $q = 0$ and $\dot{g} = 0$, there is no solution other than $\dot{s} = \infty$. The simple use of this transformation does require that u is differentiable at $t = 0$, $x = 0$ and 1, which implies that $f(0) = \phi(0)$ and $g(0) = \phi(1)$, $q(0) = \phi'(1)$ are necessary conditions. An example of a problem in which $\lambda = 0$ is given by Crank on p. 65*, and is easily put in the form (2.1) with $g = t$, $q = 0$ and $\frac{\partial u}{\partial x} = 0$ on $x = 0$. However the initial conditions $\phi(x)$ used by Crank are such that u is not differentiable at $x = 0$, $t = 0$ and the Schatz transformation will introduce a singularity there.†

The only explicit non-trivial solution in closed analytical form is that obtained by a similarity method, reducing (2.1) to a boundary value problem for an ordinary differential equation by the use of the similarity variable $\eta = x/2\sqrt{t}$. The solution requires that the free boundary is a parabola with $s(0) = 0$, and the phase does not degenerate.

It is possible to obtain some information about the time at which a phase may degenerate in the case in which $\frac{\partial u}{\partial x}$ is prescribed on $x = 0$. For example with $\frac{\partial u}{\partial x} = 0$ on $x = 0$ and $\beta = 1$, corresponding to a heat insulated boundary, the heat conservation integral (1.3) taken round the contour $x = 0$, $t = 0$, $x = s(t)$, $t = \tau$ as in Fig. 2 gives

$$\int_0^{s(\tau)} u\,dx = \int_0^1 \phi(x)\,dx + \int_0^{\tau} q(t)\,dt + \lambda(s(\tau) - 1).$$

The phase degenerates if there is a root $\tau = t_0$ of

$$\int_0^1 \phi(x)\,dx + \int_0^{\tau} q(t)\,dt = \lambda,$$

and a sufficient condition is $\int_0^{\infty} q(t)\,dt > \lambda - \bar{\phi}$ where $\bar{\phi}$ is the

---

*See also p.197 .

†See Ockendon (p.141), Fox (p.219).

mean value of $\phi(x)$ and is negative.

The one phase inverse Stefan problem is to solve (2.1) given $s(t)$ with $f(t)$ to be determined.* It is no longer a free boundary problem and is linear in u and f. The only unusual aspect in obtaining the solution is that two conditions are given on $x = s(t)$ and none on $x = 0$. An alternative simpler inverse problem is to determine $q(t)$ given s and f. A solution of an inverse Stefan problem is of course the solution to some Stefan problem and provides an indirect method for solving (2.1). Boley discusses several methods based on this approach on p.150.

The one phase Stefan problem (2.1) can be generalised in an obvious way to higher space dimensions and similar results derived to those obtained in one dimension [6].

In the original Stefan problem there are two phases and in one space dimension the boundary value problem of the first kind is

$$\left.\begin{aligned} &0<x<s(t), \quad \frac{\partial u_1}{\partial t} = \frac{\partial}{\partial x}\left(\beta_1 \frac{\partial u_1}{\partial x}\right); \\ &s(t)<x<1, \quad \frac{\partial u_2}{\partial t} = \frac{\partial}{\partial x}\left(\beta_2 \frac{\partial u_2}{\partial x}\right); \\ &x = s(t), \quad u_1 = u_2 = 0, \quad \left[\beta \frac{\partial u}{\partial x}\right]_1^2 = -\lambda \dot{s}; \\ &t = 0, \ u = \phi(x); \ x = 0, \ u_1 = f_1(t); \ x = 1, \ u_2 = f_2(t). \end{aligned}\right\} \qquad (2.3)$$

Here $\lambda, \beta_1, \beta_2$ are positive constants, $f_1 \leqslant 0$, $f_2 \geqslant 0$, $\phi \leqslant 0$ for $x < s(0) \leqslant 1$ and $\phi \geqslant 0$ for $x > s(0) \geqslant 0$. This boundary value problem is well posed [9] with a unique solution u_1, u_2, s for all t, which possesses continuous second derivatives provided ϕ, f_1 and f_2 are twice differentiable. The method of proof consists in reformulating the problem as three nonlinear Volterra integral equations and is very similar to the proof for one phase. Moreover it can be shown that this solution attains the steady state solution

$$\dot{s} = 0, \ u = (f_2(\infty) - f_1(\infty))x + f_1(\infty), \ s_\infty = \frac{f_1(\infty)}{f_2(\infty) - f_1(\infty)}$$

provided non-zero limits $f_1(\infty)$, $f_2(\infty)$ exist. The boundary value

*See Ockendon (p.143 eq(2.19)).

problem of the second kind replaces the conditions on $x = 0$ and $x = 1$ by $-\frac{\partial u_1}{\partial x} + \mu_1 u_1 = f_1(t) < 0$, $\frac{\partial u_2}{\partial x} + \mu_2 u_2 = f_2(t) > 0$ where μ_1, μ_2 are non-negative. Existence and uniqueness of the solution ($\varepsilon\ C^2$) are now only for $t < \sigma$ since one or other phase may degenerate. There are no unsteady solutions in closed analytic form except that obtained by Neumann on the interval $0 < x < \infty$ by a similarity method. With $\eta = \frac{x}{2\sqrt{t}}$, then $s = 2\alpha\sqrt{t}$, where α is a constant called the growth coefficient and is the real root of

$$2\lambda\alpha e^{\alpha^2} - f_0(\text{erf}\alpha)^{-1} - (1 - \text{erf}\alpha)^{-1} = 0, \qquad (2.4)$$

where $\phi \equiv 0 \equiv f_2$, f_1 is a negative constant f_0 and for simplicity $\beta_1 = \beta_2 = 1$. The general situation and Neumann situation are shown in Fig. 3.

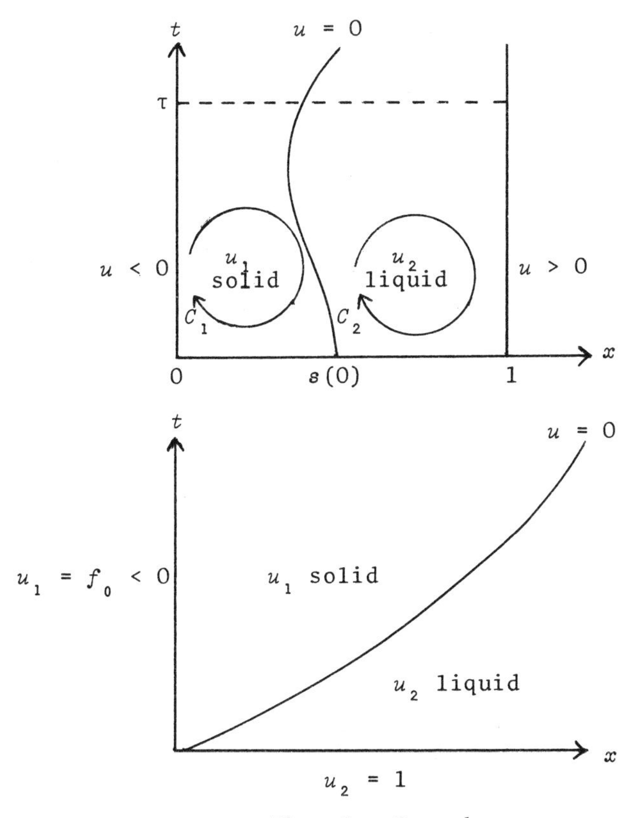

Fig. 3. Two phases.

By changing the sign of u the solid and liquid phases are interchanged. The similarity solution is still available if the convective term appears in (1.8) from a change of density between phases. The fluid velocity V is proportional to $t^{-\frac{1}{2}}$ from (1.7) and the growth coefficient satisfies a more complicated form of (2.4).

A very simple form of the two phase problem (2.3) occurs when $u_2 = 0$ on $x = 1$ for $t > 0$ and $u_2 = 0$ on $t = 0$ for $s(0) < x < 1$. Then $u_2 \equiv 0$ everywhere in $s(t) < x < 1$ and the problem reduces to a one phase problem (2.1). With more than two phases (2.3) is replaced in the obvious way by equations for $u_1, .. u_{n+1}$ and $s_1, .. s_n$. There can now be phase degeneration even for problems of the first kind and strong constraints on ϕ and f are required [2] for existence for all t. Use of the heat conservation integral taken round the contour $x = 0$, $t = 0$, $x = 1$, $t = \tau$ gives some information about possible degeneracies. For example in a three phase problem of the second kind with $\mu_1 = \mu_2 = 0$, $\beta_1 = \beta_2 = \beta_3 = 1$, we take the contour C to be the sum of the contours C_1, C_2, C_3 as in Fig. 4, where phases 1 and 3 are the same.

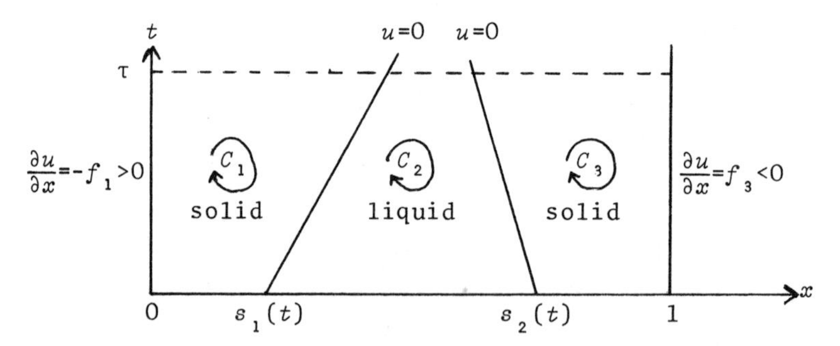

Fig. 4. Three phases.

Then

$$-\int_0^\tau f_1 dt + \int_0^1 u_{t=\tau} dx - \int_0^\tau f_3 dt - \int_0^1 \phi dx - \lambda\Big[s_1\Big]_0^\tau + \lambda\Big[s_2\Big]_0^\tau = 0.$$

Thus
$$E = \int_0^1 u_{t=\tau} dx + \lambda(s_2(\tau) - s_1(\tau)) \qquad (2.5)$$

is a known function, and when $f_1 \equiv f_3 \equiv 0$, $E(\tau) = E(0)$ is a simple "energy" integral. It is possible to show that if a solution exists for all t then

$\int_0^1 u_{t=\tau} dx \to 0$ as $\tau \to \infty$. Hence if $E(0) < 0$, $s_2 - s_1 < 0$ for some τ, and phase degenerates at $t = \sigma$ where

$$\int_0^1 u_{t=\sigma} dx = E(0).$$

If $E(0) \geqslant 0$ there may be no root σ of this equation. At times $t > \sigma$ when a phase degenerates in a multi-phase problem, u may not possess continuous second derivatives everywhere. Regions may occur in which u is identically zero, called "clouds" or "mushy regions". In higher dimensions these clouds may also occur in a two phase problem with no phase degeneracy.

To deal with this problem of lack of differentiability it is useful to introduce the concept of a weak solution* of the Stefan problem [8] and reformulate (2.3) as an integral involving no derivatives of u, rather than a differential relation. We therefore define a weak solution of a two phase Stefan problem of the first kind to be a continuous function u satisfying

$$\iint\limits_{\substack{0<x<1\\0<t<\tau}} \left\{u\frac{\partial}{\partial x}\left(\beta\frac{\partial\psi}{\partial x}\right) + A(u)\frac{\partial\psi}{\partial t}\right\}dxdt = \int_0^\tau \left[\beta f\frac{\partial\psi}{\partial x}\right]_{x=0}^{x=1} dt - \int_0^1 A(\phi)\psi_{t=0}dx \quad (2.6)$$

for any function ψ, with continuous derivatives $\frac{\partial\psi}{\partial t}$, $\frac{\partial\psi}{\partial x}$, $\frac{\partial^2\psi}{\partial x^2}$, such that $\psi = 0$ on $x = 0, 1$ and $t = \tau$, and $A(u)$ is the heat content as defined by (1.4).

It is simple to verify that any classical solution ($\in C^2$) of (2.3) also satisfies (2.6) and is therefore a weak solution. We consider

$$\iint\limits_{\substack{0<x<1\\0<t<\tau}} \left\{u\frac{\partial}{\partial x}\left(\beta\frac{\partial\psi}{\partial x}\right) + A(u)\frac{\partial\psi}{\partial t} - \psi\frac{\partial}{\partial x}\left(\beta\frac{\partial u}{\partial x}\right) + \psi\frac{\partial u}{\partial t}\right\}dxdt,$$

which may be rewritten as a contour integral using Stokes Theorem in the form

$$\int_C \left\{A(u)\psi dx + \beta\left(\psi\frac{\partial u}{\partial x} - u\frac{\partial\psi}{\partial x}\right)dt\right\}$$

*See also Atthey (p.183).

where C is the sum of contours C_1 and C_2 round the boundaries of the solid and liquid region in $0 < t < \tau$ as in Fig. 3. If u satisfies (2.3) then this integral reduces to

$$-\int_0^1 A(\phi)\psi_{t=0}dx + \int_0^\tau \left[\beta u\frac{\partial\psi}{\partial x}\right]_1^2 dt ,$$

and (2.6) is recovered.

This weak solution has two very important and possibly unexpected properties. The first is that it exists and is <u>unique</u> for all t; moreover this remains true for suitably defined weak solutions of the second kind, for multi-phase problems [7] and for higher dimensional problems [6]. The second is that certain finite difference schemes can be shown to converge to the weak solution [8] despite the discontinuities in the derivatives of u across the phase change boundary. Thus (2.6) is a powerful way of reformulating the problem both from a structural and practical point of view.

3. THE GENERALISED STEFAN PROBLEM

Many generalisations of the boundary value problem can be considered. If β is a positive function of x and t, or the boundaries $x = 0$ and $x = 1$ become non-intersecting curves, the results of the previous section are unaltered. If β is a function of u, or the boundary conditions on $x = 0$ are nonlinear, then few results are available and we shall not consider this possibility further. Other generalisations are to introduce heat source terms both in the interior and on the phase change boundary, and to allow a variable phase change temperature and latent heat. The "linear" two phase generalised Stefan problem is then

$$\left.\begin{aligned}
&0 < x < s(t), \quad \frac{\partial u_1}{\partial t} = \frac{\partial}{\partial x}\left(\beta_1(x,t)\frac{\partial u_1}{\partial x}\right) + h_1(x,t); \\
&s(t) < x < 1, \quad \frac{\partial u_2}{\partial t} = \frac{\partial}{\partial x}\left(\beta_2(x,t)\frac{\partial u_2}{\partial x}\right) + h_2(x,t); \\
&x = s(t), \quad u_1 = u_2 = g(s,t), \\
&\qquad\qquad \left[\beta\frac{\partial u}{\partial x}\right]_1^2 = -\lambda(s,t)\dot{s} - q(s,t); \\
&t = 0, \quad u = \phi(x) .
\end{aligned}\right\} \qquad (3.1)$$

In addition we are given linear conditions on $x = 0$ and 1, where all the functions β_1, β_2, h_1, h_2, g, q and λ, are differentiable and β_1, β_2 and λ are positive.

Proof of the existence and uniqueness of solutions which are twice differentiable is now not available and counter-examples in which a "mushy region" occurs are described by Atthey (p. 182) and Ockendon (p. 146). It is therefore natural to construct the weak formulation of (3.1) and this is easily done if λ is constant, $q \equiv 0$ and g is independent of s. The generalised form of (2.6) is then

$$\iint_{\substack{0<x<1 \\ 0<t<\tau}} \left\{(u-g)\frac{\partial}{\partial x}\left(\beta\frac{\partial\psi}{\partial x}\right) + A(u)\frac{\partial\psi}{\partial t} + h\psi\right\}dxdt = \int_0^\tau \left[\beta(f-g)\frac{\partial\psi}{\partial x}\right]_{x=0}^{x=1} dt - \int_0^1 A(\phi)\psi_{t=0}dx \tag{3.2}$$

where ψ is any function, defined as before, but now (1.4) becomes

$$A(u) = u \text{ for } u > g(t); \; A(u) = u - \lambda \text{ for } u \leqslant g(t).$$

In the particular case discussed by Atthey it is possible to prove the existence and uniqueness of this weak solution and devise a finite difference scheme which converges to it. In general however no results are yet available, nor is it obvious how to define a weak solution when λ and g depend s and t and q is non-zero. The corresponding generalisations of the problem to multi-phase and higher dimensional problems are obvious but few results, if any, have yet been obtained. Thus the well-posed nature of the boundary value problem (3.1) has not been established and we should therefore restrict our discussion to problems in which a physical situation is modelled and there is some experimental or intuitive evidence of a unique solution.

We therefore turn our attention to situations in which the diffusion equation is used to model the behaviour of the concentration $c(x,t)$ of some material, with non-dimensional diffusion coefficient γ taken for simplicity to be constant. Then the appropriate conservation integral corresponding to (1.3) is

$$\int_C \left(c\,dx + y\frac{\partial c}{\partial x}dt\right) = 0.$$

At a phase change boundary, where there may be discontinuities in concentration in addition to concentration gradient, application of this integral, as in (1.5), gives

$$x = s(t), \ \left[c\right]_1^2 \dot{s} + \left[\gamma \frac{\partial c}{\partial x}\right]_1^2 = 0. \tag{3.3}$$

Thus if $[c]_1^2$, on $x = s(t)$, is a given function of t, (3.3) has the form of a Stefan condition with variable coefficient λ in the notation of (1.5). In a material, such as an alloy, which consists of a pure substance together with a small concentration of a secondary substance we shall show that it is necessary to consider simultaneously the heat diffusion and concentration diffusion problems. The fundamental phenomenon is described [4] by the eutectic diagram for the material in equilibrium and such a diagram in suitably non-dimensional form is shown as Fig. 5. If the material has concentration c and temperature u such that $c > c_L(u)$ then it will be in a stable liquid phase, and if $c < c_s(u)$ it will be in a stable solid phase. Equilibrium, that is constant values of c and u, is not possible in the range $c_s(u) < c < c_L(u)$. It is however possible for the liquid and solid phase to co-exist in equilibrium at the same temperature if their respective concentrations are $c_L(u)$ and $c_s(u)$. Thus at a phase change boundary there will be a concentration discontinuity given from Fig. 5 in terms of the phase change temperature $u = g$. For small concentrations c the liquidus and solidus curves, $c = c_L(u)$ and $c = c_s(u)$, are straight lines and the ratio $k = \frac{c_s}{c_L} < 1$ is called the distribution coefficient.* For alloys with more than one secondary substance the eutectic diagram would be in a dimension higher than two.

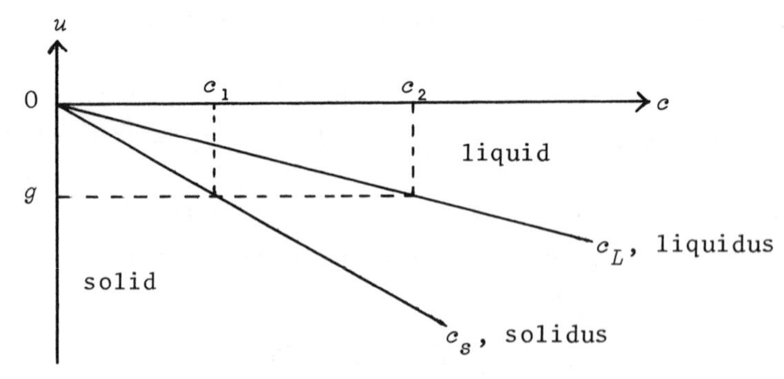

Fig. 5. Eutectic diagram.

*See Peel (p.14).

If we consider a solidification process with concentration diffusion in both phases* and assume that the eutectic diagram may be used at the moving phase change boundary then a typical boundary value problem is

$$\left.\begin{aligned}
&0 < x < s(t), \quad \gamma_1\frac{\partial^2 c_1}{\partial x^2} = \frac{\partial c_1}{\partial t}, \quad \beta_1\frac{\partial^2 u_1}{\partial x^2} = \frac{\partial u_1}{\partial t};\\
&s(t) < x < 1, \quad \gamma_2\frac{\partial^2 c_2}{\partial x^2} = \frac{\partial c_2}{\partial t}, \quad \beta_2\frac{\partial^2 u_2}{\partial x^2} = \frac{\partial u_2}{\partial t};\\
&x = s(t), \quad \left[\gamma\frac{\partial c}{\partial x}\right]_1^2 = -\left[c\right]_1^2\dot{s}, \quad \left[\beta\frac{\partial u}{\partial x}\right]_1^2 = -\lambda\dot{s},\\
&\qquad u_1 = u_2 = g, \quad c_1 = c_S(g), \quad c_2 = c_L(g);
\end{aligned}\right\} \quad (3.4)$$

together with suitable conditions on u and c, or their derivatives, at $x = 0$, 1 and $t = 0$.

The phase change temperature is now to be determined and varies with time. We would expect the problem to be well-posed but whether a mushy region exists or not ($u \in C^2$) is not clear. Similarity solutions with no mushy region do however exist on the semi-infinite range $0 < x < \infty$. The simplest occurs when no initial conditions are given, that is $-\infty < t < \infty$ when we can have a phase change boundary moving with constant speed. Then

$$\left.\begin{aligned}
s &= t, \quad c_1 = c_0, \quad u_1 = u_0 + \lambda;\\
c_2 &= c_0 - c_0\left(1 - \frac{1}{k}\right)e^{\frac{t-x}{\gamma_2}},\\
u_2 &= u_0 + (u_1 - u_0)e^{\frac{t-x}{\beta_2}}.
\end{aligned}\right\} \quad (3.5)$$

If however initial conditions $c = c_0$, $u = u_0$ are prescribed, together with boundary conditions $u \to u_0$, $c \to c_0$ as $x \to \infty$, then a similarity solution with $s = 2\alpha\sqrt{t}$ can be found if u is prescribed on $x = 0$ together with $\frac{\partial c}{\partial x} = 0$. The form of the solution is shown in Fig. 6 and is such that $\int_0^\infty (c - c_0)dx = 0$. This is not a very general solution since c is constant in the solid region but corresponds to a fairly obvious physical situation. With boundary condition $c = 1$ on $x = 0$, instead of $\frac{\partial c}{\partial x} = 0$, the physical situation is not so obvious and although the problem can be formulated in the similarity variables it is not clear that a unique real value for α exists for all c_0.

*See also Hebditch (p.112).

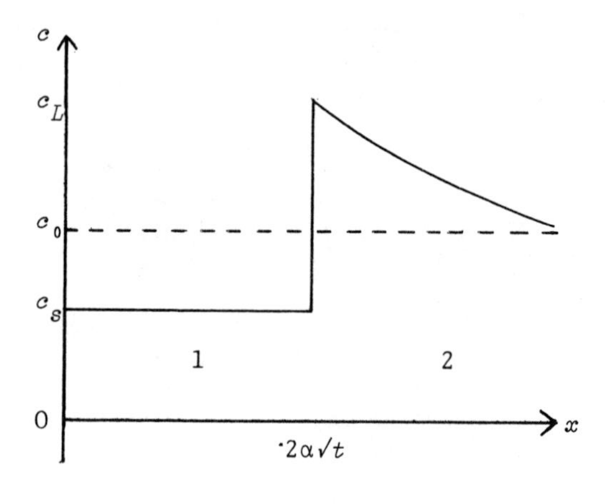

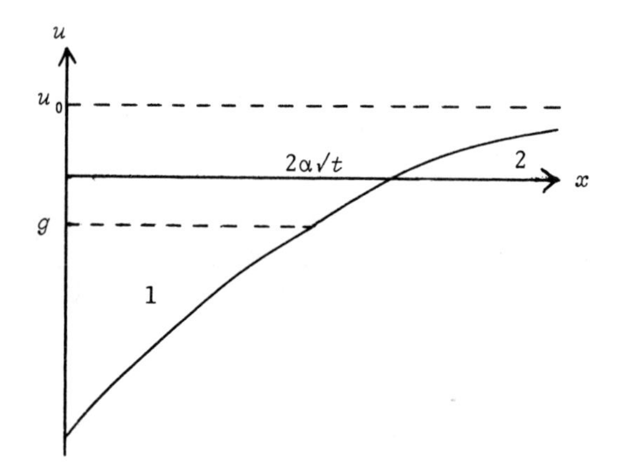

Fig. 6. Similarity solution of (3.4).

If such a value of α does exist then the concentration will vary in both phases and $\int_0^\infty (c - c_0)dx \neq 0$.

To investigate further the general problem (3.4) on the finite range, we can consider limiting values of the parameters.

(i) Instantaneous mixing: $\gamma_1 \to \gamma_2 \to \infty$.

Then $\frac{\partial^2 c_1}{\partial x^2} = 0 = \frac{\partial^2 c_2}{\partial x^2}$, with $\frac{\partial c_1}{\partial x} = \frac{\partial c_2}{\partial x}$ and $c_1 = kc_2$ at $x = s$, and given conditions on c or $\frac{\partial c}{\partial x}$ at $x = 0$ and 1. This determines c_1 and c_2 as functions of s and t and hence, from the eutectic diagram, g as a function of s and t. The problem therefore reduces to the generalised Stefan problem with $h \equiv 0$, $q \equiv 0$, λ constant, but $g = g(s,t)$. This physical situation occurs in geophysical applications and an example is discussed by Turcotte on p. 98.

(*ii*) No diffusion: $\gamma_1 \to \gamma_2 \to 0$.

Now c has to be continuous everywhere and there can be no distinct phase change boundary but a mushy region must occur corresponding to states between the liquidus and solidus lines in Fig. 5. It is not clear that this can represent a real physical situation.

(*iii*) Instantaneous heat diffusion: $\beta_1 \to \beta_2 \to \infty$.

Then $\frac{\partial^2 u_1}{\partial x^2} = 0 = \frac{\partial^2 u_2}{\partial x^2}$, with $\frac{\partial u_1}{\partial x} = \frac{\partial u_2}{\partial x}$ and $u_1 = u_2$ at $x = s$, and conditions on u at $x = 0,1$. Thus u_1 and u_2 are determined and g is given as a function of s and t. Hence c_1 and c_2 on $x = s$ are determined as functions of s and t giving a Stefan condition with the coefficient "λ" a function of s and t. This problem can be simplified even further by assuming that there is no diffusion in the solid phase so that $\gamma_1 = 0$. Then $c_1 = c_1(x)$ and $c_1(s) = c_S(g) = kc_L(g)$. Also

$$\left.\begin{aligned} &s < x < 1, \quad \gamma_2 \frac{\partial^2 c_2}{\partial x^2} = \frac{\partial c_2}{\partial t}; \\ &x = s(t), \quad c_2 = c_L(g), \quad (1-k)c_L(g)\dot{s} = -\gamma_2 \frac{\partial c_2}{\partial x}; \end{aligned}\right\} \quad (3.6)$$

with conditions on c_2 or $\frac{\partial c_2}{\partial x}$ at $x = 1$ and $t = 0$. This is a one phase Stefan problem as defined by (2.1) ($c_2 \to -u$, $x \to 1 - x$, $s \to 1 - s$) where the Stefan coefficient $(1 - k)c_L(g)$ depends on s and t, and can be shown to have a unique solution for $t < \sigma$ as before [11]. This physical situation occurs in alloy solidification and typical conditions could be $x = 1$, $\frac{\partial c_2}{\partial x} = 0$; $t = 0$, $c_2 = c_0$, $s = 0$. The form of the solution is shown in Fig. 7 and the boundary value problem (3.6) with these conditions has been used by Peel in a slightly different context on p. 15. The time σ for the phase to degenerate is of considerable physical interest.

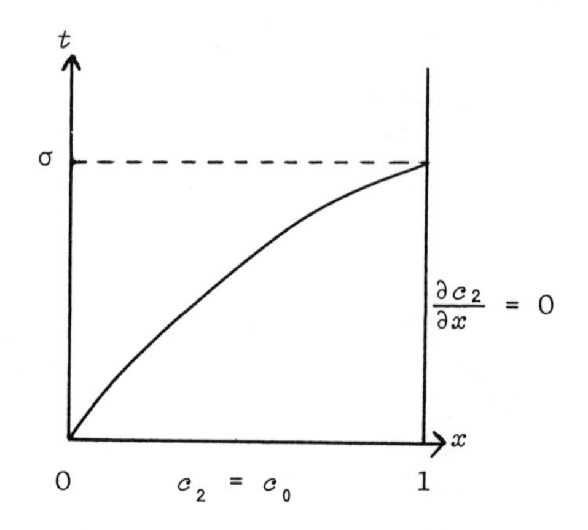

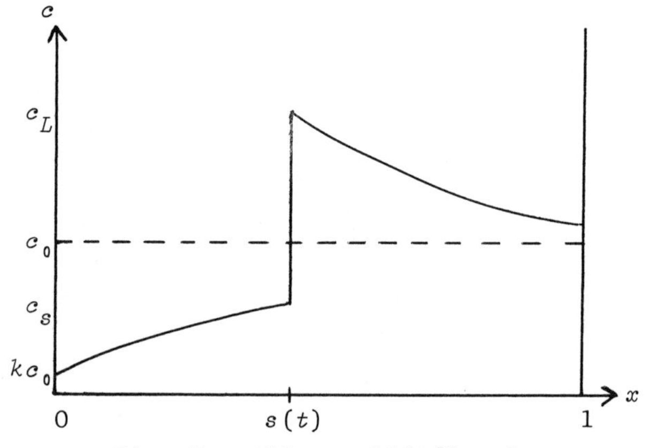

Fig. 7. Alloy solidification.

The extension of (3.4) to higher dimensions, more phases and more components is clearly relevant to real physical problems but must surely wait for further theoretical progress on (3.4) itself. The simple examples which illustrate special cases of (3.4) do provide some information about the behaviour of solutions and should provide motivation for the extension of the results obtained for the classical Stefan problem.

REFERENCES

1. Bankoff, S., "Advances in Chemical Engineering," **5**, 75-150 (1964).

2. Cannon, J., Douglas, J. and Hill, C., *J. Math. Mech.* **17**, (1967).

3. Carslaw, H. and Jaeger, J., "The Conduction of Heat in Solids," OUP., Ch.XI (1959).

4. Chalmers, B., "Principles of Solidification," Wiley (1964).

5. Crank, J., "The Mathematics of Diffusion," OUP., Ch. 7 (1967).

6. Friedmann, A., *Trans. Am. Math. Soc.* **133**, 51 (1968).

7. Friedmann, A., *Trans. Am. Math. Soc.* **133**, 89 (1968).

8. Oleinik, O., *Sov. Math. Dokl.* **1** (1960).

9. Rubinstein, L., "The Stefan Problem," AMS., 141 (1971).

10. Schatz, A., *J. Math. An. App.* **28** (1969).

11. Sherman, B., *Quart. App. Math.* **28** (1970).

12. van Moerbeke, P., *Acta Mathematica,* **132** (1974).

TECHNIQUES OF ANALYSIS

J. R. Ockendon
(University of Oxford)

1. INTRODUCTION

This paper will be concerned with several different analytical techniques for treating phase change problems involving stationary materials with constant physical properties. The classical macroscopic model of such problems as piecewise linear parabolic partial differential equations with nonlinear boundary conditions on the unknown phase boundaries has been discussed by Tayler (p.122). An alternative, more general formulation of the problem in terms of weak solutions has also been given by Tayler (p.129) and Atthey (p. 182).

Various possible reformulations of the classical model suggest themselves. For instance, a simple change of variable may fix the moving boundary [8], [16] or transform the conditions at the phase boundary into a more convenient form [28]. Although no true variational formulation exists, the quasi-variational principle of Biot [2] and some of its numerical implications are discussed by Agrawal (p. 242).

In one other important type of reformulation of the classical model, the phase boundary is shown to satisfy an integral or integro-differential equation, and the next section will be devoted to some aspects of this approach. Such a reformulation can then be used either as the point of departure for existence and uniqueness proofs, or as the basis for a numerical scheme, but we shall not discuss either of these developments here.

A mathematical tool which can often give useful information about Stefan problems is that of asymptotic analysis. Again, the variety of such applications is too great for a comprehensive survey to be given here. Instead, in Section 3, we will discuss some singular asymptotic results which suggest unexpected behaviour of the solutions of certain classes of Stefan problem.

2. INTEGRAL EQUATION FORMULATION

There is a wide variety of methods for reducing different Stefan problems to integro-differential equations which involve only conditions on the phase boundary; sometimes a bewildering array of different equations can be written down for one particular problem. There is probably most flexibility available in methods for one-phase problems in which the domain may be extended in such a way that the solution is embedded in the solution to a heat conduction problem with fixed boundaries. One such method is discussed by Boley (p. 150)*.

One standard procedure, say for the infinite two-phase problem

$$u_{1xx} = u_{1t} \quad (-\infty < x < s(t)) \tag{2.1a}$$

$$u_{2xx} = u_{2t} \quad (s(t) < x < \infty) \tag{2.1b}$$

$$u_1 = u_2 = 0, \; u_{1x} - u_{2x} = \lambda\dot{s} \text{ on } x = s(t) \tag{2.1c}$$

with u and s prescribed initially, is to define the Green's function $G(x,t) = E(\xi - x, t - \tau)$ where

$$E(x,t) = \frac{1}{2\sqrt{\pi t}} e^{-x^2/4t}. \tag{2.2}$$

Integration over the regions $x \gtrless s$ of $u(G_{xx} + G_t) - G(u_{xx} - u_t)$ and use of (2.1c) then yields three simultaneous nonlinear Volterra integral equations for $V_i(t) = u_{ix}|_{x = s(t)}$ and $s(t) - s(0) = \frac{1}{\lambda}\int_0^t \{V_1(\tau) - V_2(\tau)\}\, d\tau$. Further applications of

---

* See also [12], [24] and [31].

this method in both finite and infinite domains are given by Rubinstein [27: 94-181]†.

For problems as simple as (2.1), we may observe at once that since the solution is linear in λ and the contribution from the latent heat source on the phase boundary in time δt is $\lambda\dot{s}(t)\, E(x - s(t),t)\delta t$, the temperature distribution is given by

$$u = \phi(x) * E(x,t) + \lambda\int_0^t \dot{s}(\tau)\, E(x - s(\tau),\, t - \tau)d\tau. \quad (2.3)$$

Here ϕ is the initial temperature and $*$ denotes Fourier convolution. Then, since $u = 0$ on $x = s$, we can use the bounded convergence theorem to write down the Volterra equation

$$0 = \int_{-\infty}^{\infty} \phi(\xi)\, E(s(t) - \xi,t)d\xi + \lambda\int_0^t \dot{s}(\tau)\, E(s(t) - s(\tau),t - \tau)d\tau \quad (2.4)$$

for s. In the case of the one-phase problem when $u_2 \equiv 0$, we can also derive a Volterra equation of the second kind for $\dot{s}$ by using the fact that $u_{1x}\big|_{x=s-0} = \lambda\dot{s}$. Care now needs to be taken with the limit, since the integral in the last term of (2.3) becomes unbounded as $x \to s$, but the result is††

$$\frac{\lambda\dot{s}}{2} = \frac{\partial}{\partial x}\{\phi(x) * E(x,t)\}_{x=s} - \frac{\lambda}{2}\int_0^t \frac{\dot{s}(\tau)(s(t) - s(\tau))}{t - \tau}\,. \; E(s(t) - s(\tau),\, t - \tau)d\tau. \quad (2.5)$$

It is of interest to note that results such as (2.3) can be derived using transforms, despite the nonlinearity of the problem§. For instance, if we write

---

† Green's function techniques have also been used for example by Kolodner [14], Miranker [21], Chuang and Szekely [7] and Kruzkhov [15].

†† The idea of using heat source distributions in this way was used by Lightfoot [17]; see also [3] and [25].

§ See, for example, [19] and [22].

$$\bar{u}(k,t) = \int_{-\infty}^{s(t)} u_1 \, e^{ikx} \, dx + \int_{s(t)}^{\infty} u_2 e^{ikx} dx \qquad (2.6)$$

and note that

$$\int_{-\infty}^{\infty} u_{xx} \, e^{ikx} \, dx = \lambda \dot{s} \, e^{iks} - k^2 \bar{u} \qquad (2.7)$$

we obtain

$$\frac{\partial \bar{u}}{\partial t} + k^2 \bar{u} = \lambda \dot{s} \, e^{iks} \qquad (2.8)$$

with $\bar{u} = \bar{\phi}$ at $t = 0$. The inverse transform of the solution of (2.8) is (2.3).

Such Fourier transform techniques are less useful on finite intervals, but we note that a Laplace transform in time could also be used to obtain (2.3)*, although difficulties would arise if it was not known *a priori* that s was monotonic. When s is monotonic, however, such a procedure provides a quick way of deriving integral equations for Stefan problems on finite intervals. As an example, we consider the problem discussed by Crank (p. 165) which can be written, after the substitution $c_t = u$, as

$$u_{xx} = u_t \qquad 0 < x < s, \qquad (2.9a)$$

$$u = 0, \; u_x = -\dot{s} \qquad \text{on } x = s, \qquad (2.9b)$$

$$u = 0, \; s = 1 \qquad \text{at } t = 0, \qquad (2.9c)$$

$$u_x = \delta(t) \qquad \text{at } x = 0. \qquad (2.9d)$$

We define $s[w(x)] \equiv x$ and put $u \equiv 0$ in $x > s$, *i.e.*, $t > w$. The idea of extending the domain of definition of u to $0 < x < 1$ has already been mentioned and has been further discussed by Kruzkhov [15]. It enables us to define

$$\tilde{u} = \int_0^{w(x)} u e^{-pt} \, dt \qquad (2.10)$$

and obtain $\dfrac{d^2\tilde{u}}{dx^2} - p\tilde{u} = -e^{-pw(x)}$ (2.11a)

with $\tilde{u} = \dfrac{d\tilde{u}}{dx} = 0$ at $x = 1$ (2.11b,c)

---

* See [9].

and $\frac{d\tilde{u}}{dx} = 1$ at $x = 0$. (2.11d)

Solving (2.11a) subject to (2.11b,d), inverting and reapplying the boundary conditions at $x = s$ yields the Volterra equations derived by Hansen and Hougaard [11]. Alternatively, we may solve (2.11a,b,c) to give

$$\tilde{u} = \frac{1}{p^{\frac{1}{2}}} \int_x^1 e^{-pw(\xi)} \sinh[p^{\frac{1}{2}}(x - \xi)]\, d\xi \tag{2.12}$$

for $p > 0$, and hence obtain the Fredholm equation

$$1 = \int_0^1 e^{-pw(\xi)} \cosh(p^{\frac{1}{2}}\xi)\, d\xi \tag{2.13}$$

for w.

A similar integral formulation to (2.3) is possible for one-phase Stefan problems in which $s(t) \to \infty$ monotonically as $t \to \infty$. As an example, consider the problem

$$u_{xx} = u_t, \quad 0 < x < s, \tag{2.14a}$$

$$u = 0, \; u_x = \lambda \dot{s} \text{ on } x = s \text{ or } t = w, \tag{2.14b}$$

$$u_x = g(t) > 0 \text{ on } x = 0, \; t > 0, \tag{2.14c}$$

with $s(0) = 0$. (2.14d)

We again extend $u \equiv 0$ in $x > s$ and define

$$\tilde{u} = \int_{w(x)}^{\infty} u e^{-pt}\, dt, \tag{2.15}$$

to obtain

$$\frac{d^2\tilde{u}}{dx^2} - p\tilde{u} = -\lambda e^{-pw(x)} . \tag{2.16}$$

Using (2.14c), and assuming $\tilde{u} = o(e^{p^{\frac{1}{2}}x})$ as $x \to \infty$ gives

$$\tilde{u} = \frac{\tilde{g}(p)}{p^{\frac{1}{2}}} e^{-p^{\frac{1}{2}}x} + \frac{\lambda}{2p^{\frac{1}{2}}} \int_0^{\infty} e^{-pw(\xi)} \left[e^{-p^{\frac{1}{2}}|x - \xi|} + e^{-p^{\frac{1}{2}}(x + \xi)} \right] d\xi \tag{2.17}$$

for $p > 0$. As before, we could use the inversion of (2.17) to derive Volterra equations for w, but if we make the further assumption that $w(x)/x \to \infty$ (which is certainly true from comparison with the Neumann solution if $g(t) > O(t^{-\frac{1}{2}})$ as $t \to \infty$), then (2.15) gives $\tilde{u} = o(e^{-p^{\frac{1}{2}}x})$ as $x \to \infty$. Hence, from (2.17),

$$\frac{\tilde{g}(p)}{\lambda} = \int_0^\infty e^{-pw(\xi)} \cosh (p^{\frac{1}{2}}\xi) d\xi, \qquad (2.18)$$

a result due to Grinberg and Chekmareva [10].

The existence of such concise Fredholm equations as (2.13) and (2.18) is related to the representation of a solution of (2.14a) subject to (2.14b) in the form*

$$u = -\lambda \sum_{n=1}^{\infty} \frac{1}{(2n)!} \frac{\partial^n}{\partial t^n} [x - s(t)]^{2n}. \qquad (2.19)$$

Indeed, substituting (2.19) in (2.14c) and taking a Laplace transform in t yields (2.18).

We shall see in the next section that integral equations such as (2.18) are especially useful in obtaining asymptotic solutions.

3. ASYMPTOTIC EXPANSIONS

The dimensionless differential formulation of classical one-dimensional Stefan problems essentially involves three parameters x, t and λ and asymptotic expansions can be sought if any of these quantities is large or small. It is also possible to smooth the heat content so that it is a continuous function of temperature by defining a suitably large specific heat, say of dimensionless $O(\varepsilon^{-1})$ in the range $0 < u < \varepsilon$ occupying $s_1(t) < x < s_2(t)$. Such a device can form the basis for a numerical scheme [20] and in general the solution for small ε will tend to that of the classical Stefan problem with $|s_2 - s_1| = O(\varepsilon)$. However, $|s_2 - s_1|$ may be much larger than $O(\varepsilon)$ if the initial temperature

---

* See, for example [5].

distribution is sufficiently near zero, or, as we shall see later, if body heating is present.

Small time and distance expansions can often be written down directly from a Volterra integral formulation of the problem*. The more rarely discussed question of the stability of the phase boundary has been analysed by Wollkind & Segel [32] using conventional stability techniques.§ As far as large times and distances are concerned, several rigorous bounds can be obtained for the solution† although the precise asymptotic form of the phase boundary may not be immediately obvious. Nevertheless, for problems such as (2.14) we could derive an asymptotic differential equation for $w(x)$ by considering the solution of (2.18) as p tends to a value from whose neighbourhood the dominant contribution to the integral comes from large ξ; since $w(x) \to \infty$ as $x \to \infty$, this value is clearly $p = 0$.

Solutions for small latent heat λ are not especially interesting, there being no theory for "weak" phase changes for parabolic equations analogous to that of "weak" shock waves for hyperbolic equations. Usually, as $\lambda \to 0$ the phase boundary tends to the isotherm $u = 0$ in the solution when $\lambda = 0$, and corrections may be found by a regular perturbation expansion in λ. That this may not be the case for one-phase problems may also be seen from (2.18). Since we know that $w \to 0$ as $\lambda \to 0$, we try $w(x) \sim \omega(\lambda)x^{\alpha}$, where $\omega \to 0$ as $\lambda \to 0$. The dominant contribution to the integral in (2.18) is near $\xi = \xi_0$ where

$$-p\omega\, \alpha\, \xi_0^{\alpha-1} + p^{\frac{1}{2}} = 0 \tag{3.1}$$

and $$\frac{\tilde{g}(p)}{\lambda} = \left\{\frac{\pi}{2p\alpha(\alpha-1)\omega\, \xi_0^{\alpha-2}}\right\}^{\frac{1}{2}} e^{-p\omega\, \xi_0^{\alpha} + p^{\frac{1}{2}}\xi_0}, \tag{3.2}$$

---

* See, for example, [9], [24] and [27].

† Rubinstein [27] 155, 170; Cannon and Hill [5] and [6]; and Cannon, Douglas & Hill [4].

§ See also Peel (p.12).

Thus as $\lambda \to 0$,

$$\log \frac{1}{\lambda} \sim - p^{\frac{\alpha-2}{2(\alpha-1)}} \omega^{-\frac{1}{\alpha-1}} \left(\alpha^{\frac{-\alpha}{\alpha-1}} - \alpha^{\frac{-1}{\alpha-1}} \right) - \frac{1}{2(\alpha-1)} \log \omega, \quad (3.3)$$

and so $\alpha = 2$ and $\omega \sim \frac{1}{4 \log \lambda^{-1}}$. A rigorous derivation of this result is given by Sherman [29]. The non-uniformity of the limit $\lambda \to 0$ in one-phase problems may also be seen from the expansion of the one-phase Neumann solution (Tayler, p.127) for small λ.

The remainder of this section will be devoted to asymptotic solutions for large latent heat λ. We first consider a problem in which the timescale is sufficiently large for the phase boundary to move appreciably. The first order problem as $\lambda \to \infty$ is still nonlinear but much easier than the full problem*.

Suppose a sphere of water initially at a temperature $u = 0$ is frozen, without change of density, by imposing a temperature $u = -1$ at its surface $r = 1$. If we put $ru = v$ and replace t by λt, the problem becomes

$$v_{rr} = \frac{1}{\lambda} v_t \qquad s < r < 1, \qquad (3.4a)$$

$$v = 0,\ v_r = r\dot{s} + u = r\dot{s} \text{ on } r = s, \qquad (3.4b)$$

$$v = 0 \text{ at } t = 0 \quad r < 1, \qquad (3.4c)$$

$$v = -1 \text{ at } r = 1 \quad t > 0. \qquad (3.4d)$$

Thus, if we seek asymptotic expansions in the form

$$v \sim v_0 + \frac{1}{\lambda} v_1 + \ldots, \quad s \sim s_0 + \frac{1}{\lambda} s_1 + \ldots, \qquad (3.5)$$

we find

$$v_0 = -1 + s_0 \dot{s}_0 (r - 1) \qquad (3.6a)$$

where

$$\frac{s_0^3}{3} - \frac{s_0^2}{2} = t - \frac{1}{6}. \qquad (3.6b)$$

---

* For other examples see [18], and Andrews & Atthey (p.48).

The sphere appears to be totally frozen when $t = \frac{1}{6}$, but computation of s_1 reveals that

$$s_1 = \frac{1 - s_0}{6s_0} \tag{3.7}$$

and so the expansion (3.5) is not uniformly valid up to the time when freezing is complete. Indeed, the form of s_0 and s_1 suggests that a new expansion is required for times t such that $t_0 - t = O(\lambda^{-1})$ and distances r, s of $O(\lambda^{-\frac{1}{2}})$. This difficulty has been discussed using the method of strained coordinates by Pedroso & Domoto [23]. Recently Riley, Smith & Poots [26], by applying the method of matched asymptotic expansions, have extended the asymptotic solutions nearer to, but still not quite so far as t_0.

Moreover, the nature of the singularity (if it exists) in the temperature at such a point of phase degeneracy for smaller values of λ is still not known precisely. Similar difficulties arise in cylindrical, but not plane, geometries.

We conclude by mentioning the large latent heat solution of a problem involving body heating over a time which is short enough for the phase boundary to move only a small distance in comparison to the dimensions of the material*. Suppose ice in $0 < x < 1$ is maintained at temperature $u = -1$ at $x = 0$ and insulated at $x = 1$. A uniform body heating A is applied throughout for times $t > 0$. Then the temperature satisfies

$$u_{1xx} + A = u_{1t} \qquad (0 < x < s) \tag{3.8a}$$

$$u_{2xx} + A = u_{2t} \qquad (s < x < 1) \tag{3.8b}$$

with

$$u_1 = u_2 = 0,\ u_{1x} - u_{2x} = \lambda\dot{s} \text{ on } x = s \tag{3.8c}$$

and

$$u_1 = -1 \text{ at } x = 0,\ u_{2x} = 0 \text{ at } x = 1. \tag{3.8d}$$

---

* For other examples see [13] and [30].

We suppose the initial temperature distribution $\phi(x)$ is monotonic increasing with $\phi(1) = \phi'(1) = 0$. Initially, we seek an asymptotic solution

$$s \sim 1 + \frac{1}{\lambda} s_1(t) + \ldots . \tag{3.9}$$

In the melted region, we put $x = 1 + \frac{1}{\lambda} X$ so that

$$u_{2XX} = \frac{1}{\lambda^2}(u_{2t} - A) \tag{3.10a}$$

with

$$u_{2X} = 0 \text{ on } X = 0 \tag{3.10b}$$

and

$$u_2 = 0 \text{ on } X = s_1 \tag{3.10c}$$

Thus $u_{2x} \sim \dfrac{-AX}{2\lambda^2}$ as $\lambda \to \infty$ and so the boundary conditions for u_1 become

$$\left.\begin{aligned} u_1 &= 0 \\ u_{1x} &= \dot{s}_1 + \lambda u_{2X} \sim \dot{s}_1, \end{aligned}\right\} \text{ on } x = 1 . \tag{3.11}$$

Now the steady state solution for u_1 satisfying (3.11) is

$$u_1 = \frac{-Ax^2}{2} + \left(1 + \frac{A}{2}\right) x - 1 \tag{3.12}$$

and so, if $A > 2$, $u_1 > 0$ for x sufficiently near 1 and t sufficiently large.

This solution is only physically acceptable if we permit superheating of the ice and this situation is briefly discussed by Langham (p. 256). However, if we reject this possibility, then our model (3.8) must be modified. A study of the numerical solution of the weak formulation of the problem (Atthey, p. 183) suggests that we should allow the phase boundary to be a region, rather than a sharp curve, in the (x,t) plane. The way in which this should be done, and the implications for the asymptotic solution, are given by Atthey [1].

REFERENCES

1. Atthey, D.R., D. Phil. Thesis, Oxford University (1972).
2. Biot, M.A., *J. Aeronaut. Sci.* **24**, 857 (1957).
3. Budhia, H. and Kreith, F., *Int. J. Heat Mass Trans.*, **16**, 195 (1973).
4. Cannon, J.R., Douglas, J. and Hill, C.D. *J. Math. Mech.* **17**, 21 (1967).
5. Cannon, J.R. and Hill, C.D. *J. Math. Mech.* **17**, 1 (1967).
6. Cannon, J.R. and Hill, C.D. *J. Math. Mech.* **17**, 433 (1967).
7. Chuang, Y.K. and Szekely, J. *Int. J. Heat Mass. Trans.* **14**, 1285 (1971).
8. Crank, J. and Gupta, R.S., *Bull. Inst. Maths. and Applic.* **9**, 12 (1973).
9. Evans, G.W., Isaacson, E. and MacDonald, J.K.L., *Quart. App. Math.* **8**, 312 (1950).
10. Grinberg, G.A. and Chekmareva, O.M., *Sov. Phys. Tech. Phys.* **15**, 1579 (1971).
11. Hansen, E. and Hougaard, P. *J. Inst. Maths. and Applic.* **13** (to appear) (1974).
12. Jackson, F., *Proc. Ed. Math. Soc.* **14**, 109 (1964).
13. Jiji, L.M., *J. Franklin Inst.* **289**, 281 (1970).
14. Kolodner, I.I., *Comm. Pure App. Math.* **9**, 1 (1956).
15. Kruzkhov, S.N., *J. Appl. Math. Mech.* **31**, 1014 (1967).
16. Landau, H.G. *Quart. App. Math.* **8**, 81 (1950).
17. Lightfoot, N.M.H., *Proc. Lond. Math. Soc.* **31**, 97 (1929).
18. Lock, G.S.H., *Int. J. Heat Mass. Trans.* **14**, 642 (1970).
19. McKean, H.P., Appendix to Samuelson P.A. *Ind. Mfctg. Rev.* **6**, No. 2 (1965).
20. Meyer, G.H., *Siam. J. Num. Anal.* **10**, 522 (1973).
21. Miranker, W., *Quart. App. Math.* **16**, 121 (1958).
22. Noble, B., "Nonlinear Integral Equations" (Ed. P.M. Anselone) U. of Wisconsin Press p. 258 (1964).

23. Pedroso, R.I. and Domoto, G.A., *Int. J. Heat Mass. Trans.* 16, 1037 (1973).

24. Portnov, I.G., *Sov. Phys. Dokl.* 7, 186 (1962).

25. Rathjen, K.A. and Jiji, L.M., *J. Heat Trans.* 93, 101 (1971).

26. Riley, D., Smith, F.T. and Poots, G., *Int. J. Heat Mass. Trans.* (to appear) (1974).

27. Rubinstein, L.,"The Stefan Problem", *Trans. Math. Monographs* 27, A.M.S.

28. Schatz, A., *J. Math. An. App.* 28, 569 (1969).

29. Sherman, B., *Siam. J. Appl. Math.* 20, 319 (1971).

30. Tadjbakhsh, I. and Liniger, W., *Quart. J. Mech. Appl. Math.* 17, 141 (1964).

31. Westphal, K.O., *Int. J. Heat Mass Trans.* 10, 195 (1967).

32. Wollkind, D.J. and Segel, L.A., *Phil. Trans. Roy. Soc. London* 268A, 351 (1970).

THE EMBEDDING TECHNIQUE IN MELTING AND SOLIDIFICATION PROBLEMS*

Bruno A. Boley
(Northwestern University, Illinois)

1. INTRODUCTION

Embedding techniques represent one of the most versatile and effective methods for the solution of heat conduction problems with change of phase; for example, they can be used for the exact, approximate or numerical solution of specific problems, for the development of general analytical solutions, for the construction of upper and lower bounds, and they are equally applicable to problems in one, two or three dimensions. It will be the purpose of this survey to provide an outline of the several types of application of embedding techniques, so as to provide the reader at the same time with a basic understanding of the underlying concepts and with an overall view of their use. Most details will be omitted, with frequent reference to the literature, and particularly to the author's work.

The basic notions underlying embedding techniques are presented first, and several further developments follow, as indicated by the section headings. In addition, occasional reference will be made to results which, although not necessarily related to embedding techniques, are of general interest to workers in the field of melting and solidification.

Clearly only a very limited segment of this field is represented by the present outline; more comprehensive reviews may be found for example in [1], [4], [12] and [33].

---

* This work was supported by the Office of Naval Research.

In most of the discussion which follows we shall refer either to melting or to solidification; it is well to remember that all results apply to either case, with only obvious modifications.

2. FUNDAMENTAL NOTIONS

The fundamental concept underlying embedding techniques is simple: any one phase (restriction to two phases not being required), whose dimensions vary with time in an unknown way during the change-of-phase process, is considered to be part of a larger body, whose dimensions are identical with those of the actual body before any change of phase took place and are therefore constant in time. We speak of that particular phase as being "embedded" in a larger body of constant dimensions, although, of course, only in a mathematical sense: it is clear that only the portions of the embedding body which coincide with the actual material have any physical meanings and the others are purely fictitious. In particular, some of the boundaries of the embedding body are also fictitious: the boundary conditions there are thus not defined directly by the physical problem, and in fact they must be constructed in such a manner as to satisfy the actual conditions at the various real phase interfaces in the interior of the embedding body. How this is done will be discussed presently; first, it may be well to note the principal advantages of the technique, and thus the rationale for its adoption.

(*A*) The fact that one deals mathematically with a body of constant dimensions means that established and well known methods can be used to write the temperature explicitly in terms of arbitrary boundary and/or initial conditions.

(*B*) It follows that the boundary value problem is reduced by one spatial dimension; in particular, problems in one spatial dimension are reduced to ordinary integro-differential problems with time as the sole remaining independent variable.

These fundamental advantages permit the wide range of applications mentioned in the Introduction. It might be well, however, to point out also the principal limitations of the method.

(*a*) The methods referred to under (*A*) are normally valid only if the properties are independent of temperature. Thus embedding techniques are effective if the properties are constant in any one phase, although no difficulties arise if different phases have different properties*.

(*b*) The fact that, in general, conditions are prescribed in the interior of the embedding body means that a heat conduction problem of the "inverse" type must be solved. For long-time numerical solutions, this type of problem is known to involve questions of stability which must be handled carefully.

The fundamental relations to be satisfied at each interface express the facts that (I) the temperatures of adjacent phases must be equal to each other and to a prescribed (usually constant) temperature u_m (normally the melting or solidification temperature) and (II) an energy balance must be maintained there. Thus [22], referring to a liquid-solid interface, we have (with suffixes 1 and 2 referring to solid and liquid, respectively):

$$\text{(I)} \quad u_1(P_s;t) = u_2(P_s;t) = u_m , \tag{2.1}$$

$$\text{(II)} \quad k_1\frac{\partial u_1}{\partial n} - k_2\frac{\partial u_2}{\partial n} = \rho L v_n \tag{2.2}$$

where P_s is a point on the (moving) interface. These conditions can be expressed in a number of ways, particularly in problems involving more than one spatial dimension. A convenient general formulation was given in [31] and [33].

In the particular case in which heat is applied directly on the moving melting front, (2.2) must be replaced by

$$k_1 \frac{\partial u_1}{\partial n} = -Q+\rho L v_n . \tag{2.3}$$

---

* Note, however a limited extension to temperature dependent properties in [7].

In the notation used, k is the thermal conductivity, ρ the density, L the latent heat, n the direction of the normal to the solid-liquid interface, and v_n the velocity of motion of that interface in the n-direction. In a one-dimensional rectangular system, for example, (2.2) is simply

$$k_1\frac{\partial u_1}{\partial x} - k_2\frac{\partial u_2}{\partial x} = \rho L\frac{ds}{dt} \quad . \tag{2.4}$$

3. PROBLEMS IN ONE DIMENSION: SURFACE HEATING

The embedding technique was first introduced in [2] for the case of a slab, originally solid, and melting under the action of a prescribed surface heat flux. Two limiting cases were considered; in the first the liquid is assumed to be instantaneously removed upon formation, in the second it is assumed to remain stationary. The solution in both these cases will now be outlined; for details see [2] and [7].

Let us consider first the case in which the liquid is instantaneously removed. The solid, which originally occupied the space $0 < x < l$, is heated by a flux $Q(t)$ on $x = 0$ and is, let us say, insulated at $x = l$; clearly the conditions at the cool face $x = l$ are not crucial to an understanding of the method. The temperature rises, and first reaches u_m at $t = t_m$ on $x = 0$; thereafter a portion of thickness $s(t)$, with $s(t_m) = 0$, is melted and removed, the remaining solid occupying the space $s(t) < x < l$. The heat flux $Q(t)$ is applied on $x = s(t)$.

The embedding solid occupies at all times the space $0 < x < l$, and is acted upon, for $t > t_m$, by a fictitious heat flux on the fictitious surface $x = 0$. The solution for $t \leq t_m$ presents no difficulty, since it refers to an ordinary heat conduction problem with no moving boundaries. For $t \geq t_m$, we must write a solution for the embedding body which satisfies the following conditions.

(a) The Fourier heat conduction equation in $0 < x < l$.

(b) The condition prescribed at $x = l$.

(c) It must be identical with the pre-melting solution at $t = t_m$.

(d) It must correspond to an arbitrary heat flux $f(t)$ at $x = 0$.

These requirements define a classical heat conduction boundary value problem whose solution is most conveniently written in two parts: the first (designated by an asterisk) satisfies (a), (b), (c); the second (designated by a prime) satisfies (a), (b), (d), and homogeneous initial (*i.e.*, at $t = t_m$) conditions. Clearly then the desired temperature is

$$u(x,t) = u^*(x,t) + u'(x,t) . \tag{3.1}$$

The most convenient manner of constructing u^* is to extend the pre-melting solution into the melting regime, by continuing to apply the actual heat flux $Q(t)$ on $x = 0$, or (which is more convenient still) by applying on $x = 0$ the analytic continuation into $t > t_m$ of the pre-melting heat flux. If that is done, then a simple Taylor expansion gives (in dimensionless form)

$$\frac{u^*(x,t)}{u_m} = \sum_{n=0}^{\infty} \sum_{i=0}^{\infty} a_{ni}\, y^i\, X^{n-i} \tag{3.2a}$$

where

$$y = \frac{t-t_m}{t_m} \; ; \; X = \frac{x}{2\sqrt{K_1 t_m}} \; ; \; a_{ni} = \frac{1}{i!\,(n-i)!} \frac{\partial^n (u/u_m)}{\partial y^i \partial X^{n-i}} \tag{3.2b}$$

Here $K_1 = k_1/\rho c_1$ where c_1 is the specific heat in the solid phase.

The temperature u' is best written by means of a convolution, or Duhamel, integral. With $Q' = f - Q^*$, it is (again in dimensionless form):

$$\frac{u'(x,t)}{u_m} = \frac{1}{2}\int_0^y \frac{Q'(y-y_1)}{Q_0}\frac{\partial u_0}{\partial y_1}(X,y_1)\,dy_1 \tag{3.3}$$

where $Q_0 = \sqrt{\pi}\, k_1\, u_m/(2\sqrt{K_1 t_m})$ is a reference heat input and where u_0 is the temperature satisfying (a), (b), as well as $u_0(X,0) = 0$ and $-k_1(\partial u_0/\partial x) = 1$ on $x = 0$.

Substitution of (3.2) and (3.3) in (3.1) gives the temperature in the embedding body in terms of the single unknown quantity $Q'(y)$. This temperature satisfies all conditions of the problem, except, of course, the crucial ones to be satisfied at the unknown melting front $x = s(t)$, namely the first of (2.1) and (2.3). In other words, the two unknowns $Q'(t)$ and $s(t)$ must now be adjusted so as to satisfy the latter two equations.

These two equations define an integro-differential system [integral because of (3.3); differential because of the term $v_n = ds/dt$ in (2.3)] whose solution cannot be obtained in closed or simple analytical form. For short times, however, analytical solutions in series form are quite predictable: for example, if the heat input Q is constant, we have [2]:

$$\xi(y) = \frac{s(t)}{2\sqrt{K_1 t_m}} = \frac{2m}{3\pi}y^{\frac{3}{2}} - \frac{m^2}{4\sqrt{\pi}}y^2 - \frac{4m}{15\pi}\left(\frac{1}{2} - m^2 - \frac{16m}{3\pi^{\frac{3}{2}}}\right)y^{\frac{5}{2}} + \frac{m^2}{\sqrt{2}\sqrt{\pi}}\left(\frac{35m}{8\sqrt{\pi}} - m^2 + \frac{1}{2}\right)y^3 + \ldots \tag{3.4a}$$

where

$$m = \frac{\sqrt{\pi}}{2}\,\frac{cu_m}{L}\,. \tag{3.4b}$$

The question of the duration of the period during which series such as these may give reasonably accurate results was taken up in [29]. There, the results obtained from a few terms in the series solution are compared with the exact ones in the case of the Neumann problem for the special case of an infinite body with a spherical cavity, melting axisymmetrically outwards.

For longer times, numerical solutions must be obtained, and examples of these may be found in the analogous one-dimensional axisymmetric problems, in [28] for the cylindrical case, and in [25] for the spherical case. It is shown that, although care must be exercised in the solution of the inverse heat conduction problem which arises, the procedure is quite satisfactory.

We can now turn to the other limiting case earlier mentioned, namely that in which the liquid is allowed to remain stationary upon formation. Two phases are present after melting has begun, and it is most convenient to consider them separately. The solid phase (occupying, as in the preceding case of a melting slab the region $s(t) < x < l$), can be treated in precisely the same manner as the one in the earlier case of instantaneously removed liquid. The only difference is that, whereas the heat applied to it in the earlier problem, $Q(t)$, was prescribed and known, it is now unknown: let it be denoted by $Q_1(t)$.

The liquid phase cannot be analogously treated, because it occupies the space $0 < x < s(t)$, and the surface $x = 0$ now is exposed to a prescribed heating rate. We nevertheless still embed, mathematically, the liquid region in the slab of thickness $0 < x < l$, and note that the portion physically occupied by the liquid is initially ($t = t_m$) of zero thickness. In other words, the entire embedding body is initially fictitious, and, therefore, so is its initial temperature: hence, this may be arbitrarily prescribed. Let it be denoted by $u_m\theta(x)$, so as to maintain a dimensionless notation. The temperature in the embedding "liquid" is then easily found by classical methods, since it merely requires the temperature in a slab, under a general initial temperature $u_m\theta(x)$ and prescribed conditions at $x = 0, l$.

We can now combine liquid and solid: the temperature in each is explicitly expressed in terms of the three unknowns $Q'(t)$, $Q_1(t)$ and $\theta(x)$, and to these the fourth unknown $s(t)$ must be added. Four conditions are available for their determination, namely (2.1):

$$u_1[s(t),t] = u_2[s(t),t] = u_m \tag{3.5}$$

and (2.2), which must be split in two parts to account for the rôle played by the heat flux $Q_1(t)$ applied by the liquid to the solid, or

$$k_1\frac{\partial u_1}{\partial x} = -Q_1(t) + \rho L\frac{ds}{dt} \tag{3.6a}$$

$$k_2\frac{\partial u_2}{\partial x} = -Q_1(t). \tag{3.6b}$$

The formulation is now complete. Solution follows the same pattern as that for the case of instantaneously removed liquid, with analytical series solutions for short times and numerical ones for long times. For example, the series for the case of constant heat input has the form [2]:

$$\xi(y) = \frac{2m}{3\pi}y^{\frac{3}{2}} - \frac{m^2}{4\sqrt{\pi}}y^2 - \frac{4m}{15\pi}\left(\frac{1}{2} - m^2 - \frac{16m}{3\pi^{\frac{3}{2}}}\right)y^{\frac{5}{2}} + \\ + \left[-\frac{4Dm^3}{9\pi^2} + \frac{m^2}{12\sqrt{\pi}}\left(\frac{35m}{8\sqrt{\pi}} - m^2 + \frac{1}{2}\right)\right]y^3 + \ldots \tag{3.7}$$

where

$$D = (k_2/k_1). \tag{3.8}$$

It may be noticed that the first three terms of (3.4a) and (3.7) are identical, indicating an initial insensitivity of the solution to the manner of liquid removal. This is an important observation, since it leads to an example of the practical use of mathematical theorems on bounds and comparisons, such as are discussed in the next section.

The embedding technique can be easily extended to include problems in which internal heat generation is applied rather than surface heating [10] and [27]. It is shown in these works that no difficulty arises either conceptually or practically when the embedding formulation is adopted. In fact, the starting solution may be directly written from that of [2] for the case of a plate (of thickness $2b$; $-b < x < b$) melting under uniform and constant heat generation in a symmetric manner, so that melting starts at $t = t_m$ at $x = 0$.

It was shown in [27] that the embedding technique may also be used together with a perturbation scheme, to calculate the effect of changes in density between the liquid and the solid phases.

Among other applications of the embedding concept we may note the solution of [32] for a solid-liquid-mould three-material solution representing, for example, a problem in the solidification of castings. Since in that problem, however, the two-dimensional aspects predominate, a discussion will be postponed until Section 6.

4. BOUNDS FOR AND COMPARISONS BETWEEN SOLUTIONS

The subject of upper and lower bounds for solutions is a very large one from the mathematical standpoint, and is furthermore closely allied to the matter of general properties (e.g., uniqueness) of solution. It is not the purpose here to present a complete discussion of this topic, but only to touch upon those aspects of it which are of interest in conjunction with embedding techniques.

For purposes of illustration, let us return to the first problem discussed in Section 3, namely one in which the liquid is instantaneously removed. It is clear that corresponding to any heat function $Q'(t)$ applied to the surface $x = 0$ of the embedding body there corresponds a particular heat flux $Q(t)$ and a particular melt depth $s(t)$. These can be determined by calculating first $s(t)$ from (2.1), and $Q(t)$ from (2.3). The question then arises as to what relation, if any, the solution thus constructed on the basis of an arbitrarily chosen $Q'(t)$, bears to the desired solution for a given $Q(t)$.

The answer to this question is provided by a comparison theorem, which states [3] that if two solutions are found, corresponding, respectively, to $Q^{(1)}(t)$ and $Q^{(2)}(t)$, then

$$\left.\begin{array}{r} s^{(1)}(t) \geqslant s^{(2)}(t) \\ u^{(1)}(x,t) \geqslant u^{(2)}(x,t) \end{array}\right\} \text{if } Q^{(1)}(t) \geqslant Q^{(2)}(t). \tag{4.1}$$

Hence, if the calculated heat flux is larger than the desired one at all times, an upper bound to the solution has been constructed, and in the opposite case, a lower bound will result. It is shown in [3] that functions $Q'(t)$ can be conveniently chosen so as to obtain both types of bounds, and that furthermore these may be constructed so as to be quite close to each other. For practical purposes, then, this represents a convenient method of constructing solutions in problems of this type.

The analogous theorems and applications regarding the melting problem in which the liquid remains stationary upon formation are discussed in [5], while [4] presents a summary of the developments up to that time. One of the unsolved problems was then, as indeed pointed out in the latter reference, that of the introduction of arbitrary rates of ablation in the theorems of the type indicated above. This question was examined in [35], where it was found, among other things that a direct relationship exists between the solution and the rate of ablation. In other words, the faster the liquid is ablated or removed, the faster melting will take place.

Instantaneous liquid removal may be thought of as corresponding to an infinite rate of ablation, while obviously, if the liquid is stationary, the rate of ablation is zero. Hence, bounds for any ablation problem may be found from the comparatively simple solutions of the two limiting problems earlier described. In particular, the observation that the series solutions for these are identical up to three terms implies that these three terms are exact for *any* rate or manner of ablation. Analogous results in cylindrical and spherical coordinates are reached in [16], while the case of a material possessing several transformation temperatures was examined in detail in [34]. The practical importance of this conclusion stems in part from the fact that solutions with finite rates of ablation are extremely difficult to construct since they should include the transport equations within the moving liquid. The present discussion indicates that these complications are irrelevant for short times, and that they may be circumvented for long times by the use of bounds.

The use of upper and lower bounds for the derivation of practically useful numerical results is, to the author's mind, one of considerable importance and promise. It is nevertheless one which has received relatively little attention, and it is to be hoped not only that further research in this area will be done, but that it forms a fruitful field of co-operation between mathematicians and engineers.

5. GENERAL STARTING SOLUTIONS

As has been indicated, the embedding technique allows a reformulation of the original change-of-phase problem in a manner more convenient for the construction of a solution. In particular, the solution, examples of which for special cases have been exhibited earlier, could be derived, for short times, in general terms - depending, that is, on the character of the heating conditions before and after melting.

The general starting solution for a slab was derived in [7]. That solution gives the leading term of the series for arbitrary heating rates $Q(t)$, and includes the cases of zero and infinite rates of ablation, the effect of temperature-dependent properties, of the imposition of radiation or convection boundary conditions; detailed solutions, extending beyond the starting term, are also given there for general categories of heat inputs.

Since arbitrary rates of heating are included in the general solution, those corresponding to the Neumann problem are naturally included. In this case only, the two limiting cases do not yield identical results, and are valid only if the properties are not temperature-dependent. A method of introducing the latter effect into solutions of the Neumann type was presented in [11]. Mention may also be made of the approximate analytical solution for this type of problem introduced in [14] and [15]: it represents an extension to change-of-phase problems of the concept [13] of suitably combining *approximate* fundamental solutions so as to obtain more complex, though still approximate, solutions.

The analogous general starting solution for problems in the cylindrical coordinate system was developed in [26]. The introduction of the effect of varying densities in the liquid and in the solid, although not included in any of the works cited here, could be introduced by the perturbation approach outlined in [27].

The general solutions alluded to here have been expressed in terms of arbitrarily prescribed heating rates; they could just as easily have been written in terms of arbitrarily prescribed surface temperature. If this is done, a rather curious result is encountered [17], which will now be described. Consider the melting slab, in which the liquid remains stationary upon formation, with a surface temperature (in the earlier dimensionless notation) given by

$$u = u_m(1 + Vy^{\frac{3}{2}}), \quad t \geqslant t_m \tag{5.1}$$

where V is a dimensionless constant. Temperatures depending on powers of the square root of time are, of course, extremely common in heat conduction, and it would therefore be reasonable to expect that no difficulty would arise here, and that, in fact, the temperature (5.1) would correspond to some integral power of $\sqrt{y}$ in the supplied heat Q and in the dimensionless melt thickness $\xi(y)$. It turns out that this is not the case, but that, corresponding to (5.1), we have

$$\xi(y) = (V/\sqrt{\pi})y^{\frac{5}{4}}; \quad Q(y) = Q_0 y^{\frac{1}{4}} \tag{5.2}$$

In other words, the unexpected appearance of quarter-powers of time is encountered.

A word regarding the rôle, and indeed the importance, of short-time solutions may be in order here. Their most important characteristic is, of course, that they are exact; hence, their importance from a purely mathematical standpoint is unquestioned, as it can shed light not only on the character of the solution, but (*cf.* Section 7) on its uniqueness and in fact on its very existence. But the rôle of short-time solutions is often an essential one from a practical standpoint as well, as a few examples will show.

In the first place, it has been shown [29] that in certain cases the validity of the short-time solution extends over a considerable range, and also permits some theoretical discussion of the prevalent practice of employing Neumann's solution in terms of an empirically determined parameter. More important, the initial stages of, for example, casting solidification are of considerable metallurgical importance, and cannot be accurately studied by approximate or numerical methods because of the singularities arising at the instant at which a change of phase initiates; a discussion of this may be found, for example, in [32]. In general, however, numerical solutions are found to be extremely sensitive to changes in accuracy at short times (e.g. [23], [25] and [28]), and it is therefore often convenient - and indeed, essential - to start the solution analytically, and then to continue it numerically. At the least, exact short-time solutions are useful in providing a much needed check on both numerical and approximate analytical solutions.

The last point is particularly important if the temperatures obtained are to be subsequently employed in further analyses, notable among these being those connected with the determination of the mechanical behaviour (*i.e.*, stresses, strains and deformations) of the melting or solidifying body. This is true because errors in the temperature may be magnified in the corresponding errors in the mechanical variables. In fact, it was found necessary in certain cases (e.g. [24]) to modify the use of the concept of penetration depth (used in conjunction with the heat-balance method) in such a manner as to obtain an asymptotically exact starting solution.

6. PROBLEMS IN MORE THAN ONE SPATIAL DIMENSION

The first application of the embedding technique to two-dimensional problems is that of [33], where several problems were solved in which the heating conditions vary along the surface. Although obviously more complicated than the one-dimensional ones previously discussed, these solutions are conceptually similar to them, and simply involve an additional series expansion to describe the variation along the surface.

The direct extension of one-dimensional ideas to multi-dimensional situations is possible only in certain types of problems, which were identified in [33], and are intimately connected with the conditions at the start of melting. Three cases were in fact distinguished: (I) melting starting simultaneously at all points of the surface; (II) melting starting simultaneously at all points of a portion of the surface; (III) melting starting at a point of the surface. It is clear that in problems of Classes II and III spread of the melting interface will occur both into the body and along the surface, while the latter will be absent in problems of

Class I. It is this difference which renders problems of Class II and III essentially different from those of Class I (which includes one-dimensional ones, incidentally) and materially more difficult.

The work cited above introduced the above classification and, except for some illustrative examples, restricted itself to problems of Class I. The same is true of [32], where the not inconsiderable additional complications encountered stemmed not from this source, but from the simultaneous consideration of three heat-conducting media separated by imperfectly conducting contacts. It should, of course, be mentioned that approximations to problems of Classes II and III can at times be obtained by introducing in Class I suitably sharp surface temperature variations.

One of the results obtained in [30] may be quoted here as being of special practical significance. This refers to the fact that sharp variations in local heat transfer (caused, for example, by imperfect mould contact in a casting) are not reflected in a similarly sharp progress of the solidification front. Indeed, the latter appears to be much more "one-dimensional" than the corresponding two-dimensional disturbance, a fact that is both known and important to metallurgical applications.

The solution of some problems of Class III (that is, with melting of the half-space $z > 0$ starting at the origin and spreading both in the z-direction and along the plane $z = 0$) represents probably the most significant basic advance made possible by the embedding technique. The first solution was that of [20] and referred to the plane problem (in the x,z-plane). It was found by employing a number of assumptions about the distribution of the fictitious surface heat input and of the melt front location; these quantities were expressed in series form, and the coefficients of the series were numerically calculated from an infinite set of linear simultaneous equations. The convergence of these numerical calculations was rather slow; nevertheless, some important conclusions were reached, among them the facts that melting would progress much faster along the surface than towards the interior (respectively, proportionally to $\tau^{\frac{1}{2}}$ and to $\tau^{\frac{3}{2}}$, where τ is dimensionless time), and that for short times the melting interface would be a universal function, independent, that is, of the details of the surface heating distribution.

Examination of a plot of the universal function just alluded to indicated a good fit by the equations

$$\xi(X,\tau) = A_1\tau^{\frac{3}{2}}\left[1 - \left(\frac{X}{X_0(\tau)}\right)^2\right]^{\frac{3}{2}} \tag{6.1}$$

$$X_0(\tau) = A\tau^{\frac{1}{2}} \tag{6.2}$$

with suitably assigned numerical values of the constants A and A_1. In these equations $\tau = (t - t_m)/t_m$, *i.e.*, the quantity previously indicated by y, and all coordinates and distances are rendered dimensionless by dividing by $(2\sqrt{K_1 t_m})$. Although no theoretical grounds existed for the assumption that these relations might represent the exact solutions, an attempt was made [36] to introduce them as initial assumptions, and to see whether the constants A and A_1 could be determined so as to satisfy all conditions of the problem. Remarkably, this turned out to be the case; not only that, but the analogous radially symmetric problem proved to be equally amenable to treatment. With this encouragement, the conjecture was made that even the general three-dimensional problem of a half-space melting from a surface point could be similarly solved on the basis of the assumptions:

$$\xi(X,Y,\tau) = A_2\tau^{\frac{3}{2}}\left[1 - \left(\frac{X}{X_0(\tau)}\right)^2 - \left(\frac{Y}{Y_0(\tau)}\right)^2\right]^{\frac{3}{2}} \tag{6.3}$$

where

$$\left(\frac{X}{X_0}\right)^2 + \left(\frac{Y}{Y_0}\right)^2 \leqslant 1$$

and

$$X_0(\tau) = A\tau^{\frac{1}{2}}\ ;\ Y_0(\tau) = B\tau^{\frac{1}{2}}\quad . \tag{6.4}$$

The validity of this conjecture was established in [21], where the general solution was found, and where the constants A, B and A_2 were shown to be:

$$A = \left(2\frac{C_1}{C_{2x}}\right)^{\frac{1}{2}}\left[1 - \frac{1}{4}\left(\frac{1}{2}\frac{C_{2x}}{C_1}\right) + \frac{11}{32}\left(\frac{1}{2}\frac{C_{2x}}{C_1}\right)^2 + \frac{1}{8}\left(\frac{1}{2}\frac{C_{2x}}{C_1}\right)\left(\frac{1}{2}\frac{C_{2y}}{C_1}\right) + \ldots\right] \tag{6.5}$$

$$B = 2\left(\frac{C_1}{C_{2y}}\right)^{\frac{1}{2}}\left[1 - \frac{1}{4}\left(\frac{1}{2}\frac{C_{2y}}{C_1}\right) + \frac{11}{32}\left(\frac{1}{2}\frac{C_{2y}}{C_1}\right)^2 + \frac{1}{8}\left(\frac{1}{2}\frac{C_{2x}}{C_1}\right)\left(\frac{1}{2}\frac{C_{2y}}{C_1}\right) + O\left(\frac{C_{2x}}{C_1} + \frac{C_{2y}}{C_1}\right)^3\right] \tag{6.6}$$

$$A_2 = \frac{4}{3}\frac{m}{\pi}\left[1 + \frac{1}{4}\left(\frac{1}{2}\frac{C_{2x}}{C_1}\right) + \frac{1}{4}\left(\frac{1}{2}\frac{C_{2y}}{C_1}\right) - \frac{3}{32}\left(\frac{1}{2}\frac{C_{2x}}{C_1}\right)^2 - \frac{3}{32}\left(\frac{1}{2}\frac{C_{2y}}{C_1}\right)^2 - \frac{1}{16}\left(\frac{1}{2}\frac{C_{2x}}{C_1}\right)\left(\frac{1}{2}\frac{C_{2y}}{C_1}\right) + O\left(\frac{C_{2x}}{C_1} + \frac{C_{2y}}{C_1}\right)^3\right] \tag{6.7}$$

where

$$C_1 = \frac{\partial}{\partial\tau}\left(\frac{u}{u_m}\right) \; ; \quad C_{2x} = \frac{\partial^2}{\partial X^2}\left(\frac{u}{u_m}\right) \; ; \quad C_{2y} = -\frac{\partial^2}{\partial Y^2}\left(\frac{u}{u_m}\right)$$

all evaluated at $X = Y = Z = \tau = 0$.

7. UNIQUENESS OF SOLUTION

The three-dimensional solutions just described were obtained by means of a semi-inverse procedure, namely one in which certain portions of the solution were assumed and others were determined in the course of the development. Such a procedure is not satisfactory unless accompanied by a uniqueness theorem, that is to say, one which would insure that no other initial assumption would have been possible. Such a theorem derived in [6], showed that a sufficient condition for uniqueness was that the applied heat flux should be constant in direction and dependent on time only. These conditions were satisfied in all the previous solutions, and no difficulties therefore arose.

Uniqueness of solution for the one-dimensional case in which heat was applied at an arbitrary rate $Q(t)$ was assured by the work of [3], [4] and [5]. The more general case of a heat input dependent on both time and position was examined in [8], and led to the conclusion that conditions sufficient for uniqueness were (a) that Q be independent of position, or (b) that it decrease monotonically with increasing distance into the melting body, or (c) that it satisfy certain conditions of Lipschitz continuity. The analogous development for the three-dimensional case, a generalisation of that of [6], was presented in [9].

In order to gain a better understanding of the conditions under which uniqueness was assured, and possibly to develop some counter examples for the cases not covered by these conditions, the general solution of [7], referred to in Section 5, was employed to construct some general classes of solutions. This work as described in [8], led to some interesting observations regarding both uniqueness and existence of solution in cases of instantaneous liquid removal under space- and time-dependent heat inputs.

The work of [7] shows that (for example) the most general heat input which will yield a melting depth proportional to $y^{\frac{3}{2}}$ (y indicating, as earlier, dimensionless time), *i.e.*,

$$\xi(y) = cy^{\frac{3}{2}} \tag{7.1}$$

is of the form

$$\frac{Q(y)}{Q_{.0}} = \sum_n a_n \xi^n y^{(1-3n)/2} \tag{7.2}$$

for any constants c, a_n, and n, among which a certain relationship must hold. Several special cases were examined; in particular, for $n = 2$, it was found that for $a_2 \leqslant 0$, a unique solution (*i.e.*, a single value of c) would exist, but that for positive values of a_2 either two values of c would satisfy all conditions of the problem or, if a_2 were sufficiently large, no solution at all would exist. This indicates that care must be exercised in problems of this type that they are physically well posed; in other words, although melting cannot occur here in the manner which has been mathematically stipulated, it may still occur in a different manner. The rather unrealistic nature of the heat inputs required for non-existence of solution should also be pointed out; nevertheless, this appears to be a suitable area for further inquiry.

8. MECHANICAL BEHAVIOUR IN THE PRESENCE OF PHASE CHANGE

Some work in this area has already been referred to (*i.e.*, [24]); more recent work in the area of the melting of nuclear reactor fuel plates was reported in [10], and further references may be found for example in [4] and [12]. These works do not rely on the embedding technique, and are therefore somewhat beyond the scope of the present brief review. Nevertheless, they refer to an area of paramount practical importance and high theoretical interest, but unfortunately comparatively little explored: for this reason, the author felt that some mention of it would not be inappropriate, and might encourage others to enter it.

It should be mentioned that the difficulties offered by any analysis of mechanical behaviour are great, not the least of them being introduced by the uncertainty of the appropriate type of material behaviour at the very large temperatures involved. For example, [12] and the earlier [19] assumed an elastic-perfectly plastic behaviour, with a yield stress linearly decreasing from its value at room temperature to zero at the melting temperature, but scant information is available on this score. Experimental results are sorely needed; in fact, it might be fitting to end by emphasising the need for, and the importance of, more and more accurate experimental results in the entire field of both thermal and mechanical behaviour of bodies undergoing changes of phase.

REFERENCES

1. Bankoff, S.G., "Heat Conduction or Diffusion with Change of Phase", *Advances in Chemical Engineering*, **5** (editors: T. B. Drew, J. H. Hoopes and T. Vermeulen), Academic Press, New York, 75-150 (1964).

2. Boley, B.A., "A Method of Heat Conduction Analysis of Melting and Solidification Problems", *Journal of Math. and Physics*, **40**, No.4, 300-313 (1961).

3. Boley, B.A., "Upper and Lower Bounds for the Solution of a Melting Problem", *Quarterly of Applied Mathematics*, **21**, No. 1, 1-11 (1963).

4. Boley, B.A., "The Analysis of Problems of Heat Conduction and Melting", *High Temperature Structures and Materials*, proceedings of III Symposium on Naval Structural Mechanics, Pergamon Press, New York, 260-315 (1964).

5. Boley, B.A., "Upper and Lower Bounds in Problems of Melting or Solidifying Slabs", *Quarterly Journal of Mechanics and Applied Mathematics*, **17**, Part 3, 253-269 (1964).

6. Boley, B.A., "Estimate of Errors in Approximate Temperature and Thermal Stress Calculation", proceedings XI International Congress of Applied Mechanics. Springer Verlag (1964).

7. Boley, B.A., "A General Starting Solution for Melting or Solidifying Slabs", *International Journal of Engineering Science*, **6**, 89-111 (1968).

8. Boley, B.A., "Uniqueness in a Melting Slab with Space- and Time- Dependent Heating", *Quarterly of Applied Mathematics*, **27**, No. 4, 481-487 (1970).

9. Boley, B.A., "Uniqueness of Solution in Three-Dimensional Melting and Solidification Problems",*Tech. Report No. 1*, Department of Theoretical and Applied Mechanics, Cornell University (1970).

10. Boley, B.A., "Temperature and Deformations in Rods and Plates Melting under Internal Heat Generation",

proceedings of First International Conference on Structural Mechanics in Reactor Technology , compiled by T. A. Jaeger, Berlin, **6**, Part L, Paper L2/3 (1971).

11. Boley, B.A., "On a Melting Problem with Temperature-Dependent Properties", *in* Trends in Elasticity and Thermoelasticity, W. Nowacki Anniversary Volume, Wolters-Noordhoff Publishing, Groningen, The Netherlands, 22-28 (1971),

12. Boley, B.A., "Survey of Recent Developments in the Fields of Heat Conduction in Solids and Thermoelasticity", *Nuclear Engineering and Design*, **18**, No. 3, 377-399 (1972).

13. Boley, B.A., "On the Use of Superposition in the Approximate Solution of Heat Conduction Problems", *International Journal of Heat and Mass Transfer*, **16**, 2035-2041 (1973).

14. Boley, B.A., "Applied Thermoelasticity", *Developments in Mechanics*, **7**, 3-16 (1973).

15. Boley, B.A., "Methods of Approximate Heat Conduction Analysis", proceedings of Second International Conference on Structural Mechanics in Reactor Technology , compiled by T. A. Jaeger, Berlin, **5**, Part L, Paper L1/2 (1973).

16. Boley, B.A., Grimado, P.B. and Lederman, J.M., "Radially Symmetric Melting of Cylinders and Spheres", proceedings IV International Heat Transfer Conference, Versailles, (1970).

17. Boley, B.A. and Lee, Y.F., "A Problem in Heat Conduction of Melting Plates", *Letters in Applied and Engineering Sciences*, **1**, 25-32 (1973).

18. Boley, B.A. and Weiner, J.H., "Theory of Thermal Stresses", John Wiley and Sons, New York (1960).

19. Boley, B.A. and Weiner, J.H. "Elasto-Plastic Thermal Stresses in a Solidifying Slab", *Journal of Mechanics and Physics of Solids*, **11**, 145-154 (1963).

20. Boley, B.A. and Yagoda, H.P., "The Starting Solution for Two-Dimensional Heat Conduction Problems with Change of Phase", *Quarterly of Applied Mathematics*, **27**, No. 2, 223-246 (1969).

21. Boley, B.A. and Yagoda, H.P., "The Three-Dimensional Starting Solution for a Melting Slab", Proc. Roy. Soc. London A. 89-110 (1971).

22. Carslaw, H.S. and Jaeger, J.C., "Conduction of Heat in Solids", Clarendon Press, Oxford, Second Edition (1959).

23. Citron, S.J., "On the Conduction of Heat in a Melting Slab", proceedings IV U.S. National Congress of Applied Mechanics (1962).

24. Friedman, E. and Boley, B.A., "Stresses and Deformations in Melting Plates", *Journal of Spacecraft and Rockets*, **7**, No. 3, 324-333 (1970).

25. Grimado, P. and Boley, B.A., "A Numerical Solution for the Symmetric Melting of Spheres", *International Journal of Numerical Methods in Engineering*, **2**, 175-188 (1970).

26. Guzelsu, A.N. and Cakmak, A.S., "Starting Solution for an Ablating Hollow Cylinder", *International Journal of Solids and Structures*, **6**, 1087 (1970).

27. Lahoud, A. and Boley, B.A., "Some Considerations of the Melting of Reactor Fuel Plates and Rods", *Nuclear Engineering and Design*, *1974*; presented at Second International Conference on Structural Mechanics in Reactor Technology, Berlin 1973 (1974).

28. Lederman, J.M. and Boley, B.A., "Axisymmetric Melting or Solidification of Circular Cylinders", *International Journal of Heat and Mass Transfer*, **13**, 413-427 (1970).

29. Lee, Y.F. and Boley, B.A., "Melting an Infinite Solid with a Spherical Cavity", *International Journal of Engineering Science*, **11**, No. 12, 1277 (1973).

30. Muehlbauer, J.C. and Sunderland, J.E., "Heat Conduction with Freezing or Melting", *Applied Mechanics Reviews*, **18**, 951-959 (1965).

31. Patel, P.D., "Interface Conditions in Heat Conduction Problems with Change of Phase", *AIAA Journal*, **6**, 2454 (1968).

32. Patel, P.D. and Boley, B.A., "Solidification Problems with Space and Time Varying Boundary Conditions and Imperfect Mold Contact", *International Journal of Engineering Science*, **7**, 1041-1066 (1969).

33. Sikarskie, D.L. and Boley, B.A., "The Solution of a Class of Two-Dimensional Melting and Solidification Problem", *International Journal of Solids and Structures*, **1**, No. 2, 207-234 (1965).

34. Wu, T.S., "Bounds in Melting Slab with Several Transformation Temperatures", *Quarterly Journal of Mechanics and Applied Mathematics*, **XIX**, Part 2, 183-195 (1966).

35. Wu, T.S. and Boley, B.A., "Bounds in Melting Problems with Arbitrary Rates of Liquid Removal", *SIAM Journal Applied Mathematics*, **14**, No. 2, 306-323 (1966).

36. Yagoda, H.P. and Boley, B.A., "Starting Solutions for Melting of a Slab under Plane or Axisymmetric Hot Spots", *Quarterly Journal of Mechanics and Applied Mathematics*, **XXIII**, Part 2 (1970).

THE SOLUTION OF A TWO-PHASE STEFAN PROBLEM BY A VARIATIONAL INEQUALITY

G. Duvaut
(University of Paris)

1. THE PHYSICAL PROBLEM

We consider a problem involving solid-liquid phase changes at a prescribed temperature. The temperature at the phase boundary is taken to be zero, with $u < 0$ ($u > 0$) corresponding to the solid (liquid) state. The physical applications which we have in mind in considering such problems are the following.

(*i*) Melting or solidification of ice or water in a reservoir, pipe or a porous medium. (We neglect density changes in the reservoir, which is assumed to be slightly elastic, and calculate these density changes separately *a posteriori*. We may thus suppose that the two phases are at rest.)

(*ii*) Problems in metallurgy involving melting of a metal in a crucible or solidification of a weld. This type of problem can be exemplified by the melting of a block of ice which is initially at temperature $u_0(\underset{\sim}{x}) < 0$ contained in a fixed region Ω. Density changes are ignored and it is assumed that the melted ice remains at rest. For times $t > 0$ a section Γ_1 of the boundary $\partial\Omega$ is at a temperature $\hat{b}(t)$ which for the sake of simplicity is supposed constant for $\underset{\sim}{x} \in \Gamma_1$. Across the complement, Γ_2, of Γ_1 in $\partial\Omega$, there is a given heat flux which is taken to be zero in this example. The block of ice will melt subject to these conditions and the melting surface $L(t)$ will separate the solid and liquid states. If the initial temperature $u_0(\underset{\sim}{x})$ is zero and $\hat{b}(t)$ is positive, the region Ω separates into one region where the temperature is zero and another region where it is strictly positive. When $\hat{b}(t)$ is constant, this "one-phase" Stefan problem (see Tayler p.123) can be reduced to a variational inequality [2].

In the work which follows, however, the initial temperature can be different from zero and in general the temperatures are not constant in either of the two phases. We will also assume that the coefficients of thermal conductivity and the specific heats in the water and the ice may take different constant values in the two phases.

In Sections 2 to 4 the equations for the problem are stated and a variational formulation of the problem is derived. In Section 5 it is shown that this variational problem possesses a unique solution and the way it can be used to obtain the numerical solution of the problem is described in Section 6.

2. STATEMENT OF THE PROBLEM

Let Ω be the open region occupied by the block of ice and $\partial\Omega$ its boundary consisting of Γ_1 and Γ_2 (Fig. 1).

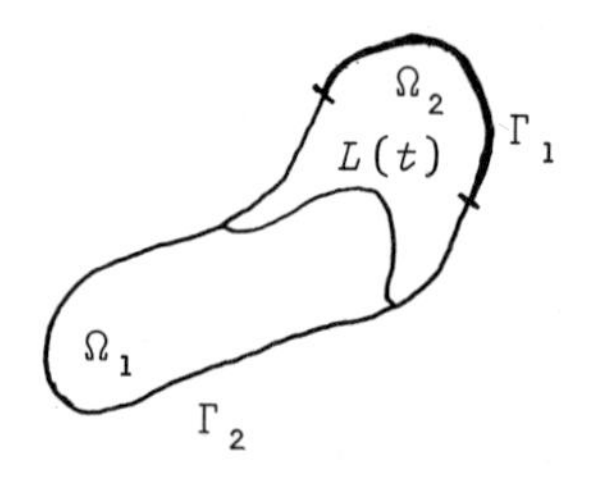

Fig. 1.

The temperature field is $u(\underset{\sim}{x},t)$, where $\underset{\sim}{x} = (x_1,x_2,x_3)$ and the time $t \in (0,\tau)$ ($\tau > 0$, given), and $L(t)$ is the phase boundary. This surface $L(t)$ divides Ω into two regions Ω_1 and Ω_2 representing, respectively, the solid and the liquid phases. We will suppose initially that the temperature in the interior of Ω_1 is strictly negative and in the interior of Ω_2 is strictly positive. Having formulated this problem mathematically we will then verify that the equations remain valid even if there are regions of zero temperature within Ω_1 or Ω_2.

The heat conduction equation for material of unit density is

$$c_i \frac{\partial u}{\partial t} - k_i \nabla^2 u = 0 \text{ in } \overset{0}{\Omega}_i \tag{2.1}$$

where $\overset{0}{\Omega}_i$ is the interior of Ω_i, and the constants c_i and k_i are, respectively, the specific heat and the coefficient of thermal conductivity in $\overset{0}{\Omega}_i$. If $t = \ell(\underset{\sim}{x})$ is the equation of the surface $L(t)$ then

$$u(\underset{\sim}{x},t) = 0 \text{ on } L(t). \tag{2.2}$$

Moreover if $\underset{\sim}{n}$ is the unit normal to $L(t)$ directed towards Ω_1, if V_n is the speed of $L(t)$ along $\underset{\sim}{n}$, and if L is latent heat of fusion of ice, we have

$$LV_n = \underset{\sim}{q}_2 . \underset{\sim}{n} - \underset{\sim}{q}_1 . \underset{\sim}{n} \tag{2.3}$$

where

$$\underset{\sim}{q}_i = - k_i \text{ grad } u \tag{2.4}$$

is the heat flux in Ω_i.
Thus

$$(k_2 \text{ (grad } u)_2 - k_1 \text{ (grad } u)_1).\text{grad } \ell = - L \tag{2.5}$$

The boundary conditions are

$$u(\underset{\sim}{x},t) = \hat{b}(t) \text{ on } \Gamma_1 \tag{2.6}$$

$$- \frac{\partial u}{\partial n} = 0 \text{ on } \Gamma_2 \tag{2.7}$$

and the initial condition is

$$u(\underset{\sim}{x},0) = u_0(\underset{\sim}{x}) \text{ in } \Omega. \tag{2.8}$$

3. CHANGE OF DEPENDENT VARIABLE

Let $\underset{\sim}{x}$ be a point in Ω. Initially this point is in the solid phase. Let $t_1(\underset{\sim}{x})$ be the first time that $\underset{\sim}{x}$ is in the liquid phase, and let $t_2(\underset{\sim}{x}) > t_1(\underset{\sim}{x})$ be the next time at which $\underset{\sim}{x}$ is again in the solid phase. Thus we may define a sequence

$$t_0(= 0),\ t_1,\ t_2,\ \ldots..$$

such that

$$\begin{array}{l}\text{in } (t_i, t_{i+1}),\ \underset{\sim}{x} \text{ is in the liquid} \\ \text{(solid) phase if } i \text{ is odd (even).}\end{array} \tag{3.1}$$

We now introduce the new unknown function $v(\underset{\sim}{x},t)$ defined by

$$v(\underset{\sim}{x},t) = k_1\int_{t_0}^{t_1} u(\underset{\sim}{x},\tau)d\tau + k_2\int_{t_1}^{t_2} u(\underset{\sim}{x},\tau)d\tau + \ldots k_j\int_{t_i}^{t} u(\underset{\sim}{x},\tau)d\tau \quad (3.2)$$

where $t \varepsilon (t_i, t_{i+1})$ and $j = 1(2)$ if i is even (odd). Alternatively

$$v(\underset{\sim}{x},t) = \int_0^t (-k_1 u^-(\underset{\sim}{x},\tau) + k_2 u^+(\underset{\sim}{x},\tau))d\tau$$

$$\text{where } u^+ = \sup(u,0),\ u^- = \sup(-u,0). \quad (3.3)$$

Since u is continuous in $\overset{0}{\Omega}$, from (3.2) we obtain

$$\frac{\partial v}{\partial t} = k_j\ u(\underset{\sim}{x},t) \quad (3.4)$$

where $j = 1(2)$ if i is even (odd) for $t \varepsilon (t_i,\ t_{i+1})$.

Differentiating equation (3.2) we obtain, since $u(\underset{\sim}{x},t_i) = 0$,

$$\text{grad } v = k_1\int_0^{t_1} \text{grad } u(\underset{\sim}{x},\tau)d\tau + \ldots + k_j\int_{t_i}^{t} \text{grad } u(\underset{\sim}{x},\tau)d\tau. \quad (3.5)$$

Then, differentiating a second time, we find

$$\nabla^2 v = k_1\int_0^{t_1} \nabla^2 u(\underset{\sim}{x},\tau)d\tau + \ldots + k_j\int_{t_i}^{t} \nabla^2 u(\underset{\sim}{x},\tau)d\tau$$

$$+ k_1(\text{grad } u(\underset{\sim}{x},t_1))_1 .\ \text{grad } t_1 - k_2(\text{grad } u(\underset{\sim}{x},t_1))_2 .\ \text{grad } t_1$$

$$+ \ \ldots\ldots - k_j(\text{grad } u(\underset{\sim}{x},t_i))_j .\ \text{grad } t_i. \quad (3.6)$$

Using (2.1) and (2.5) this becomes

$$\nabla^2 v = \int_0^{t_1} c_1\frac{\partial u}{\partial \tau}(\underset{\sim}{x},\tau)d\tau + \ldots + \int_{t_i}^{t} c_j\frac{\partial u}{\partial \tau}(\underset{\sim}{x},\tau)d\tau + L - L \ldots, \quad (3.7)$$

which gives

$$\nabla^2 v = -\ c_1\ u(\underset{\sim}{x},0) + L + \frac{c_2}{k_2}\frac{\partial v}{\partial t} \quad \text{if} \quad u(\underset{\sim}{x},t) > 0$$

$$\text{and } \nabla^2 v = -\ c_1\ u(\underset{\sim}{x},0) + \frac{c_1}{k_1}\frac{\partial v}{\partial t} \quad \text{if} \quad u(\underset{\sim}{x},t) < 0. \quad (3.8)$$

Since u has the same sign as $\frac{\partial v}{\partial t}$, (3.8) can also be written

$$\nabla^2 v = -\ c_1\ u(\underset{\sim}{x},0) + \frac{c_2}{k_2}\left(\frac{\partial v}{\partial t}\right)^+ - \frac{c_1}{k_1}\left(\frac{\partial v}{\partial t}\right)^- + LH\left(\frac{\partial v}{\partial t}\right), \quad (3.9)$$

where

$$H(\xi) = 0 \quad \text{if} \quad \xi < 0$$
$$H(\xi) = 1 \quad \text{if} \quad \xi > 0. \tag{3.10}$$

Equation (3.9) is defined except at points where $u(\underset{\sim}{x},t) = 0$. Let us look more closely at what happens at such points. There are several possibilities.

(*i*) The only place where $u(\underset{\sim}{x},t) = 0$ is on the phase boundary $L(t)$. This surface divides Ω into two regions Ω_1 and Ω_2 where the temperatures are strictly negative and positive respectively. Equation (3.9) is then well-defined in both Ω_1 and Ω_2. It has no meaning on $L(t)$ which has zero measure in Ω.

(*ii*) At time t, there exists in Ω_1 (solid phase) an interior non-empty region where $u(\underset{\sim}{x},t) = 0$. At all points $\underset{\sim}{x}$ inside this region, $\frac{\partial u}{\partial t}$ and $\nabla^2 u$ are zero since (2.1) is satisfied. Suppose $t \in (t_i, t_{i+1})$ at the point $\underset{\sim}{x}$. Then $u(\underset{\sim}{x},t) = 0$ for all $t \in (t_i, t_{i+1})$; this follows since $u(\underset{\sim}{x},t) \leqslant 0$ (we are in the solid phase throughout $(t_i,\ t_{i+1})$) and u is an analytic function of t throughout the interval. Consequently

$$v(\underset{\sim}{x},t) = k_1 \int_0^{t_1} u(\underset{\sim}{x},\tau)\,d\tau + \ldots + k_2 \int_{t_{i-1}}^{t_i} u(\underset{\sim}{x},\tau)\,d\tau \tag{3.11}$$

which implies that

$$\frac{\partial v}{\partial t} = 0, \quad \text{grad} v = k_1 \int_0^{t_1} \text{grad} u(\underset{\sim}{x},\tau)\,d\tau + \ldots + k_2 \int_{t_{i-1}}^{t_i} \text{grad} u(\underset{\sim}{x},\tau)\,d\tau \tag{3.12}$$

and that

$$\begin{aligned} \nabla^2 v = {} & k_1 \int_0^{t_1} \nabla^2\ u(\underset{\sim}{x},\tau)\,d\tau + \ldots + k_2 \int_{t_{i-1}}^{t_i} \nabla^2\ u(\underset{\sim}{x},\tau)\,d\tau \\ & + (k_1\ (\text{grad } u)_1 - k_2 \text{ grad } u)_2).\ \text{grad } t_1 + \ldots. \\ & + (k_1\ (\text{grad } u)_1 - k_2 \text{ grad } u)_2).\ \text{grad } t_{i-1} \\ & + \ k_2\ (\text{grad } u)_2.\ \text{grad } t_i\ . \end{aligned} \tag{3.13}$$

Now, at t_i, $(\text{grad } u)_1 = 0$, and so

$$k_2\ (\text{grad } u)_2.\ \text{grad } t_i = -\ L \tag{3.14}$$

which implies that

$$\nabla^2 v = k_1 \int_0^{t_1} \nabla^2 u(\underset{\sim}{x},\tau)\,d\tau + \ldots + k_2 \int_{t_{i-1}}^{t_i} \nabla^2 u(\underset{\sim}{x},\tau)\,d\tau$$

$$= \int_0^{t_1} c_1 \frac{\partial u}{\partial \tau}(\underset{\sim}{x},\tau)\,d\tau + \ldots + \int_{t_{i-1}}^{t_i} c_2 \frac{\partial u}{\partial \tau}(\underset{\sim}{x},\tau)\,d\tau$$

$$= - c_1\, u(\underset{\sim}{x},0). \tag{3.15}$$

Thus equation (3.9) is still valid if we now define $H\left(\frac{\partial v}{\partial t}\right) = 0$ in the solid region.

(*iii*) Similarly it can be shown that if there is a region in the liquid phase where $u(x,t) = 0$ we must take $H\left(\frac{\partial v}{\partial t}\right) = 1$ when $\frac{\partial v}{\partial t} = 0$.

Thus we may in any case write

$$H\left(\frac{\partial v}{\partial t}\right) \in \partial\phi_0 \left(\frac{\partial v}{\partial t}\right) \tag{3.16}$$

where $\phi_0(x) = x^+$ (*cf.* (3.3)) and the subgradient notation

$$f(x) \in \partial\phi(x) \iff \phi(\xi) - \phi(x) \geqslant (\xi - x)\, f(x)$$

for all $\xi \in \mathbb{R}$.

The boundary conditions to be satisfied by the function $v(\underset{\sim}{x},t)$ are

$$v(\underset{\sim}{x},t) = k_1 \int_0^{t_1} \hat{b}(\tau)\,d\tau + \ldots. + k_j \int_{t_i}^{t} \hat{b}(\tau)\,d\tau$$

$$= \int_0^t (k_2(\hat{b}(\tau))^+ d\tau - k_1(\hat{b}(\tau))^- d\tau \text{ on } \Gamma_1, \tag{3.17}$$

$$\frac{\partial v}{\partial n} = 0 \quad \text{on} \quad \Gamma_2 \tag{3.18}$$

and the initial condition is

$$u(\underset{\sim}{x},0) = 0. \tag{3.19}$$

SUMMARY

If we define the convex function $\phi_1 : \mathbb{R} \to \mathbb{R}$ by

$$\phi_1(x) = \tfrac{1}{2} \frac{c_1}{k_1} (x^-)^2 + \tfrac{1}{2} \frac{c_2}{k_2} (x^+)^2 + Lx^+, \tag{3.20}$$

then

$$\nabla^2 v + c_1 \; u(\underset{\sim}{x},0) \; \varepsilon \; \partial\phi_1 \left(\frac{\partial v}{\partial t}\right) \qquad (3.21)$$

which is equivalent to the statement that

$$\phi_1(\xi) - \phi_1\left(\frac{\partial v}{\partial t}\right) \geqslant (\nabla^2 v + c_1 \; u(\underset{\sim}{x},0))(\xi - \frac{\partial v}{\partial t}) \text{ for all } \xi \; \varepsilon \; \mathbb{R}. \qquad (3.22)$$

We therefore need to find $v(\underset{\sim}{x},t)$ such that:

(*i*) v, grad$v$, $\frac{\partial v}{\partial t}$ are continuous in Ω, (3.23)

(*ii*) v satisfies (3.22) in Ω, (3.24)

(*iii*) v satisfies the boundary conditions (3.17) and (3.18) and the initial condition (3.19). (3.25)

Conversely for a given $v(\underset{\sim}{x},t)$, we can state that $\frac{\partial v}{\partial t}(\underset{\sim}{x},t) > 0$ (<0) implies that $\underset{\sim}{x} \; \varepsilon \; \Omega_2$ (Ω_1). If, however, $\frac{\partial v}{\partial t}(\underset{\sim}{x},t) = 0$ in an open region of Ω, then $\nabla^2 v(\underset{\sim}{x},t) + c_1 \; u(\underset{\sim}{x},0) = 0$ ($=L$) implies $\underset{\sim}{x} \; \varepsilon \; \Omega_1$ (Ω_2). When the temperature field $u(\underset{\sim}{x},t)$ is given, then $v(\underset{\sim}{x},t)$ is given by (3.4), namely $\frac{\partial v}{\partial t} = k_2 \; u^+ - k_1 \; u^-$.

4. VARIATIONAL FORMULATION

We begin by defining the space of admissible functions

$$U(t) = \{\omega | \omega \, \varepsilon \, H^1(\Omega), \omega|_{\Gamma_1} = b(t)\} \qquad (4.1)$$

where H^1 is the Sobolev space of functions whose generalised first derivatives are square-integrable, and

$$b(t) = k_2(\hat{b}(t))^+ - k_1(\hat{b}(t))^-. \qquad (4.2)$$

For $\omega \; \varepsilon \; U(t)$ we substitute $\xi = \omega(\underset{\sim}{x})$ in (3.22) and integrate over Ω to obtain

$$\Phi(\omega) - \Phi\left(\frac{\partial v}{\partial t}\right) \geqslant \int_\Omega \nabla^2 v(\omega - \frac{\partial v}{\partial t}) d\underset{\sim}{x} + c_1 \int_\Omega u(x,0)(\omega - \frac{\partial v}{\partial t}) d\underset{\sim}{x} \qquad (4.3)$$

where

$$\Phi(\omega) = \int_\Omega \phi_1(\omega) d\underset{\sim}{x}. \qquad (4.4)$$

We transform (4.3) by Green's Theorem to obtain

$$a(v,\omega - \frac{\partial v}{\partial t}) + \Phi(\omega) - \Phi\left(\frac{\partial v}{\partial t}\right) \geqslant c_1 \int_\Omega u(\underset{\sim}{x},0)(\omega - \frac{\partial v}{\partial t}) d\underset{\sim}{x} \qquad (4.5)$$

where

$$a(v,\omega) = \int_\Omega \text{grad}v.\ \text{grad}\omega\ d\underset{\sim}{x}.$$

Furthermore

$$\frac{\partial v}{\partial t} \in U(t) \tag{4.6}$$

and

$$v(\underset{\sim}{x},0) = 0. \tag{4.7}$$

We can now formulate the Variational Principle: If $v(\underset{\sim}{x},t)$ is a solution of (3.23) to (3.25) then $v(\underset{\sim}{x},t)$ satisfies (4.4), (4.6) and (4.7).

5. EXISTENCE AND UNIQUENESS RESULTS

Theorem

Given positive constants c_i, k_i $(i = 1,2)$, L and functions $u_0(\underset{\sim}{x}) = u(\underset{\sim}{x},0)$ and $b(t)$ with

$$u_0 \in L^2(\Omega), \quad b \in L^2(0,T), \quad (T > 0 \text{ given}), \tag{5.1}$$

define the convex functions ϕ_1 by (3.20), Φ by (4.4) and the bilinear functional $a(v,\omega)$ by (4.5). Then there exists a unique $v(\underset{\sim}{x},t)$ such that:

(*i*) $v,\ \frac{\partial v}{\partial t} \in L^2(H^1(\Omega);\ 0,T)$, (5.2)

(*ii*) $\frac{\partial v}{\partial t} \in U(t)$, (defined by (4.1))

(*iii*) $a(v,\omega - \frac{\partial v}{\partial t}) + \Phi(\omega) - \Phi\frac{\partial v}{\partial t} \geqslant c_1\int_\Omega u_0(\underset{\sim}{x})(\omega - \frac{\partial v}{\partial t})\,d\underset{\sim}{x}$
for all $\omega \in U(t)$ and all $t \in (0,T)$, (5.3)

(*iv*) $v = 0$ when $t = 0$.

For a proof of this theorem we refer the reader to work on monotone parabolic inequalities by Brésis [1].

6. OUTLINE OF NUMERICAL SCHEME

We use a finite difference scheme in the time variable and let h (>0) be the step length. We shall show that if $v(t)$ is supposed known, then $v(t + h)$ may be calculated by minimising a convex functional.

$$\text{We approximate } \frac{\partial v}{\partial t} \text{ by } \frac{v(t+h) - v(t)}{h}. \tag{6.1}$$

Since $\frac{\partial v}{\partial t} \in U(t)$ for all t, we have

$$\frac{v(t+h)}{h} \in \hat{U}(t) = \{\hat{\omega} \mid \hat{\omega} = \omega + \frac{v(t)}{h}, \quad \omega \in U(t)\}. \tag{6.2}$$

We introduce the convex functional $\hat{\Phi}(\hat{\omega})$ by

$$\hat{\Phi}(\hat{\omega}) = \Phi(\omega) = \Phi\left(\hat{\omega} - \frac{v(t)}{h}\right). \tag{6.3}$$

Then, inequality (5.3) can be written as

$$\begin{aligned} &a\left(v(t), \omega - \frac{v(t+h)}{h}\right) + \hat{\Phi}(\hat{\omega}) - \hat{\Phi}\left(\frac{v(t+h)}{h}\right) \\ &\geqslant c_1 \int_\Omega u_0(\underset{\sim}{x}) \left(\hat{\omega} - \frac{v(t+h)}{h}\right) d\underset{\sim}{x} \text{ for all } \hat{\omega} \in \hat{U}(t). \end{aligned} \tag{6.4}$$

Thus the unknown $\frac{v(t+h)}{h}$ minimises the functional

$$I(\hat{\omega}) = \hat{\Phi}(\hat{\omega}) - c_1 \int_\Omega u_0(x)\hat{\omega}\, d\underset{\sim}{x} + a(v(t), \hat{\omega}) \tag{6.5}$$

on the closed convex space $U(t)$. It can be shown from the theory of minimisation of convex functionals [3] that the minimum is indeed attained in the space $\hat{U}(t)$. Thus it is possible to use classical numerical methods for the minimisation of convex functionals to obtain $v(t+h)$ at each timestep.

REFERENCES

1. Brézis, H., Comptes rendus Ac. Sc. Paris. 274A, 340 (1972). See also Attouch, H., Benilan, Ph., Damlamian, A., Picard, C., "Equations d'évolution avec conditions unilatérales," C.R. Acad. Sciences, Paris, to appear.
2. Duvaut, G., "Résolution d'un problème de Stéfan," C.R. Acad. Sc. Paris, 276, 1461-1463 (1973). See also Duvaut, G., (1973) New variational techniques in mathematical physics, Bressanone, Italy.
3. Lions, J.L., "Contrôle optimal des systèmes gouvernés par des équations aux dérivées partielles, "Dunod, Paris (1968).

A Finite Difference Scheme for Melting Problems Based on the Method of Weak Solutions

D. R. Atthey

(C.E.G.B., Marchwood Engineering Laboratories)

1. INTRODUCTION

We shall consider the problem of heat conduction in an idealised material with constant thermal properties in which a phase change occurs, such as melting or solidification. This phase change is assumed to take place at some specified temperature. Such a problem is frequently referred to as a Stefan problem, after early work by Stefan [7]. The usual treatment of a Stefan problem assumes the existence of some moving surface such that the material is in its solid phase on one side of the surface and in its liquid phase on the other side. Energy balance arguments then show that although the temperature is continuous everywhere, the heat flux has a jump discontinuity across the phase change surface due to the absorption or liberation of latent heat at this surface. For example, in a one dimensional problem in which we have solid material for $x < X(t)$ and liquid for $x > X(t)$, this discontinuity is

$$\left[-k\partial T/\partial x\right]_{\text{solid}}^{\text{liquid}} = \rho L dX/dt \qquad (1.1)$$

where T is the temperature, k the thermal conductivity, ρ the density and L the latent heat per unit mass. The nonlinearity of this jump condition is the principal cause of difficulty in any analytical or numerical treatment of the problem.

When a volume heat production term is added to the heat conduction equation, the melting or solidification may take place in a different manner from that described above. It is possible that there will no longer be a sharp boundary between the liquid and the solid, but rather a "mushy region" in which the material is at its melting temperature and the solid and liquid phases coexist. An example of such a situation is given by Atthey [1].* With Stefan problems in several space dimensions, a further complication may arise. Even though the data for the problem may be perfectly smooth, the solution need not be. For example, a single block of ice which is melting may at some time break up into several smaller islands. Thus the "free boundary" between the solid and liquid regions varies in a discontinuous manner.

In view of the various physical complications at the phase boundary, we are led to search for a weak formulation of the problem in which it is not necessary to describe explicitly the behaviour of the solution near any of its singularities. The formulation given below follows that of Oleinik [6] and Kamenomostskaja [4].

2. WEAK SOLUTIONS†

For simplicity we first consider a heat conduction problem in which no phase change is involved. We consider a volume G which is bounded by a surface ∂G and in which there is body heating $A(\underset{\sim}{r},t,T)$ per unit volume, from the time $t = 0$ to $t = t_0$. We consider only the boundary condition

$$T(\underset{\sim}{r},t) = g(\underset{\sim}{r},t) \text{ on } \partial G, \tag{2.1}$$

although other boundary conditions could be treated in a similar manner. An energy balance over an arbitrary volume V with boundary ∂V which is contained in the volume G shows that

$$\underbrace{\int_{\partial V} k\underset{\sim}{\nabla}T.d\underset{\sim}{S}}_{\substack{\text{(rate of flux of heat}\\ \text{across surface)}}} + \underbrace{\int_V A\,dV}_{\substack{\text{(rate of heat generation}\\ \text{within volume)}}} = \underbrace{\frac{\partial}{\partial t}\int_V \rho H\,dV}_{\substack{\text{(rate of increase}\\ \text{of heat content)}}} \tag{2.2}$$

*See also Ockendon (p.147), Langham (p.256).

†See also Tayler (p.129).

where H is the heat content per unit mass. H is a function of the temperature T, and if no phase change is involved and the specific heat, c, is constant we may set

$$H = cT . \tag{2.3}$$

As the solution of the heat conduction problem contains no singularities we may use Green's theorem to yield

$$\int_V \{k\nabla^2 T + A - \rho\partial H/\partial t\} dV = 0 . \tag{2.4}$$

By summing equations of the form (2.4) over different volumes V, and integrating with respect to time, we may show that

$$\int_0^{t_0}\int_G \{k\nabla^2 T + A - \rho\partial H/\partial t\}\phi dV dt = 0 \tag{2.5}$$

for a piecewise constant function $\phi(\underset{\sim}{r},t)$. By considering uniformly convergent sequences of piecewise constant functions, it follows that (2.5) holds for an arbitrary continuous function ϕ. We now restrict our attention to *test functions* ϕ, *i.e.*, to functions ϕ which satisfy the conditions

$$\phi(\underset{\sim}{r},t) = 0 \quad \text{on } \partial G \tag{2.6}$$

$$\phi(\underset{\sim}{r},t_0) = 0 \quad \text{in } G \tag{2.7}$$

and for which the second space derivatives and first time derivative are continuous. If the initial condition to our heat conduction problem has the form

$$H(\underset{\sim}{r},0) = h_0(\underset{\sim}{r}) \qquad \text{in } G , \tag{2.8}$$

we may integrate (2.5) by parts and use (2.1), (2.6), (2.7) and (2.8) to yield

$$\int_0^{t_0}\int_G \{kT\nabla^2\phi + A\phi + \rho H\partial\phi/\partial t\} dV dt$$
$$= -\int_G \rho h_0(\underset{\sim}{r})\phi(\underset{\sim}{r},0)dV + \int_0^{t_0}\int_{\partial G} kg(\underset{\sim}{r},t)\nabla\phi . d\underset{\sim}{S} dt . \tag{2.9}$$

In the derivation of the integral equation (2.9) we have made use of the fact that for a heat conduction problem with no change of phase, the higher order derivatives of the functions T and H exist and are continuous. This is not the case for a heat conduction problem which has a phase change, as across the phase change surface, $\underset{\sim}{\nabla} T$ will be discontinuous. We shall, however, make use of (2.9) in our definition of a weak solution. For a material which undergoes a change of phase at temperature $T = 0$ with a corresponding latent heat L, the relationship between the temperature T and the heat content H becomes

$$H = \begin{cases} cT & : T < 0 \\ cT + L & : T > 0 \,. \end{cases} \tag{2.10}$$

We now *define* a weak solution of a heat conduction problem with a change of phase, subject to the boundary condition (2.1) and the initial condition (2.8) to be a pair of bounded functions T and H related by (2.10) for which the integral equation (2.9) is satisfied for all test functions ϕ. We shall now show that the above definition of a weak solution includes the usual jump condition across a phase change surface. For simplicity we consider the one dimensional form of the problem.

If we replace the boundary condition (2.1) by

$$\begin{aligned} T &= g_1(t) \text{ on } x = l_1 \\ T &= g_2(t) \text{ on } x = l_2 \end{aligned} \tag{2.11}$$

the one dimensional version of (2.9) becomes

$$\begin{aligned} &\int_0^{t_0}\int_{l_1}^{l_2} \{kT\partial^2\phi/\partial x^2 + A\phi + \rho H\partial\phi/\partial t\}dxdt \\ &= -\int_{l_1}^{l_2} \rho h_0(x)\phi(x,0)dx - \int_0^{t_0} kg_1(t)\partial\phi(l_1,t)/\partial x\, dt \\ &\quad + \int_0^{t_0} kg_2(t)\partial\phi(l_2,t)/\partial x\, dt. \end{aligned} \tag{2.12}$$

We assume that there are two regions R_1 and R_2 which are separated by a curve $x = X(t)$ (Fig. 1) and in each of which

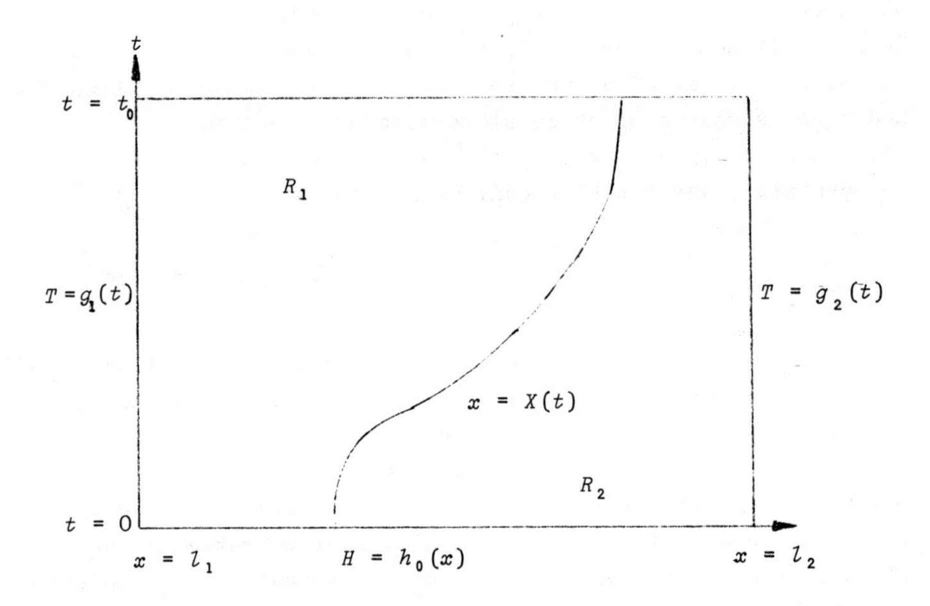

Fig. 1 The x, t plane for a weak solution containing jump discontinuities in H and $\partial T/\partial x$.

the one dimensional form of the classical heat conduction equation holds. Multiplying the heat conduction equation by the test function ϕ, integrating over R_1 and integrating by parts we have

$$\iint_{R_1} \{kT\partial^2\phi/\partial x^2 + A\phi + \rho H\partial\phi/\partial t\}dxdt$$

$$= -\int_{l_1}^{X(0)} \rho h_0(x)\phi(x,0)dx - \int_0^{t_0} kg_1(t)\partial\phi(l_1,t)/\partial x dt$$

$$+ \int_{x = X(t) - 0} \{(kT\partial\phi/\partial x - k\phi\partial T/\partial x)dt - \rho H\phi dx\}. \tag{2.13}$$

A similar equation holds in R_2. Adding these two, subtracting (2.12) and assuming that T is continuous across $x = X(t)$, we have

$$\int_{x = X(t)} \phi\{k[\partial T/\partial x]dt + [H]\rho dx\} = 0, \qquad (2.14)$$

where $[\cdot]$ denotes the jump discontinuity from $x = X(t) - 0$ to $x = X(t) + 0$. (2.14) holds for an arbitrary test function ϕ and an arbitrary time t_0, so

$$dX/dt = [-k\partial T/\partial x]/\rho[H] \qquad (2.15)$$

If the region R_1 contains solid material and the region R_2 contains liquid material, then $[H] = L$ and (2.15) becomes the usual jump condition, *i.e.*, (1.1).

We have thus shown that both the classical solution of a heat conduction problem in several dimensions, but with no change of phase, and the classical solution to a one dimensional Stefan problem satisfy the criterion for a weak solution. The extensions to the cases of Stefan problems in a more complicated geometry or with variable thermal properties are straightforward and will not be included here. An important result is that a heat conduction problem (with or without a change of phase) has a unique weak solution [6]. Thus if we investigate a heat conduction problem for which we know that a classical solution exists, any weak solution that we may find will also be the (unique) classical solution. This is in contrast to the situation arising in the study of hyperbolic equations, say the equations of gas dynamics, where further restrictions are needed to render the similarly defined weak solutions unique.

The principal advantage gained by considering the weak solution rather than the classical solution to a Stefan problem lies in the fact that derivatives of the temperature do not appear in the weak formulation of the problem. Thus it is no longer necessary to consider the liquid and solid regions separately. Instead, the entire region of interest is considered, and the shape of the liquid and solid regions may be determined later from the nature of the solution.*
Moreover, several authors including Oleinik [6], Kamenomostskaja

*See also Eggleton (p.108).

[4] and Friedman [3] have found it convenient to use the concept of a weak solution to discuss the question of existence and uniqueness of solutions to Stefan problems.

Motivated by the above discussion, we shall now describe a finite difference scheme for phase change problems in one dimension.

3. THE FINITE DIFFERENCE SCHEME

We consider the solution of a one dimensional melting problem with body heating in a material with constant specific heat. The differential equation for heat conduction is

$$k\partial^2 T/\partial x^2 + A = \rho \partial H/\partial t \ . \qquad (3.1)$$

We deliberately do not pose a melting condition such as (1.1), but merely define $T = T(H)$ from (2.10) by

$$\begin{aligned} T &= H/c \quad &, H \leqslant 0 \\ T &= 0 \quad &, 0 \leqslant H \leqslant L \\ T &= (H-L)/c \quad &, H \geqslant L \ . \end{aligned} \qquad (3.2)$$

We shall consider only the boundary conditions

$$\begin{aligned} T &= g_1(t) \quad \text{on} \quad x = l_1 \\ T &= g_2(t) \quad \text{on} \quad x = l_2 \end{aligned} \qquad (3.3)$$

although other boundary conditions could be treated similarly. We take as the initial boundary

$$H(x,0) = h_0(x). \qquad (3.4)$$

We now consider the solution of the problem (3.1) - (3.4) by an explicit finite difference scheme. We choose $t_0 > 0$, and for positive integers M and N we put $\delta x = (l_2 - l_1)/M$, $\delta t = t_0/N$. The finite difference scheme is:

$$H_m^{n+1} = H_m^n + (k\lambda/\rho)\,(T_{m+1}^n - 2T_m^n + T_{m-1}^n) + A(T_m^n)\delta t/\rho \qquad (3.5)$$

where T^n_m is an approximation to the temperature at the point $x = l_1 + m\delta x$, $t = n\delta t$ etc., and $\lambda = (\delta t/\delta x^2)$. The boundary conditions are

$$\left.\begin{aligned} T^n_0 &= g_1(n\delta t) \\ T^n_1 &= g_2(n\delta t) \end{aligned}\right\} \quad n = 0,\ 1,\ 2,\ \ldots N, \tag{3.6}$$

and the initial condition is

$$H^0_m = h_0(l_1 + m\delta x), \qquad m = 0,\ 1,\ 2,\ \ldots\ M. \tag{3.7}$$

We define piecewise constant functions $\bar{T}$ and $\bar{H}$ on $[l_1, l_2]$ x $[0, t_1]$ by

$$\bar{T}(x,t) = T^n_m \tag{3.8}$$

$$\bar{H}(x,t) = H^n_m \tag{3.9}$$

for $(m - 1)\delta x \leqslant x < m\delta x$, $n\delta t \leqslant t < (n + 1)\delta t$.

Because of the expected discontinuities in H and $\partial T/\partial x$, the usual proofs for the convergence of the scheme to a solution of the heat conduction equation [2] do not apply to a problem with a change of phase. However, it is possible to prove that as M and $N \to \infty$, the functions T and H converge to a weak solution of our heat conduction problem. The existence of a weak solution follows as an immediate consequence of this proof of convergence. A proof of the convergence of the scheme to a weak solution is given by Kamenomostskaja [4] or Atthey [1].

Our reason for replacing the classical formulation of the Stefan problem by the modified heat conduction equation (3.1) and the enthalpy equation (3.2) is now clear. The weak solution to which our numerical solution converges may contain singularities of the type described at the end of Section 2, but the behaviour of the solution near these singularities does not now have to be specified *a priori*. Certainly with the classical Stefan problem in one dimension this leads to considerable programming simplifications. A further advantage is obtained in problems where the nature of the solution near its singularities is less easy to specify in advance, such as that of a material which is being melted by a volume distribution of heat sources.

Although a large number of numerical methods have been developed for the Stefan problem in one dimension,* very few of these may be generalised to cover Stefan problems in several space dimensions. However, the extension of the above finite difference scheme to problems in more than one dimension is straightforward, and the proof of convergence follows that of the one dimensional case.

Essential to the proof of convergence of this finite difference scheme is the restriction on the steplengths:

$$\lambda = (\delta t/\delta x^2) \leqslant \lambda_0 < (\rho c/2k) \ . \tag{3.10}$$

This is the usual condition for the stability of the solution to a parabolic equation by an *explicit* finite difference method [2]. For certain problems it is possible that the restriction imposed on the time step δt by (3.10) will lead to a prohibitively large number of computations. For such problems, it would be preferable to attempt a solution by an *implicit* finite difference scheme, which would not be subject to a restriction such as (3.10). If (3.5) is replaced by a finite difference formula involving temperatures at the time steps $t = n\delta t$ and $t = (n + 1)\delta t$, then use of the enthalpy equation (3.2) will yield a system of nonlinear equations for the unknown H_m^{n+1}. An alternative approach is that of Meyer [5] who approximates the enthalpy equation (3.2) by taking H as a continuous, piecewise linear function of T, with the effect of the latent heat spread over some temperature range ε. An implicit finite difference scheme yields a system of nonlinear equations for the temperature T_m^{n+1}, which may be solved by an iterative method.

---

*See Fox (p.210).

REFERENCES

1. Atthey, D. R., *J. Inst. Maths. Applics.*, **13**, 353-366 (1974).
2. Fox, L., "Numerical solution of ordinary and partial differential equations", Oxford, Pergamon Press, 230 (1962).
3. Friedman, A., *Trans. Am. Math. Soc.*, **133**, 51-87 and 89-114 (1968).
4. Kamenomostskaja, S. L., *Mat. Sb.*, **53** (95), 489-514 (1961).
5. Meyer, G. H., *SIAM J. Numer. Anal.*, **10**, 522-538 (1973).
6. Oleinik, O. A., *Soviet Math. Dokl.*, **1**, 1350-1354 (1960).
7. Stefan, J., *S-B Wien Akad. Mat. Natur*, **98**, 173 (1889).

FINITE-DIFFERENCE METHODS

John Crank
(Brunel University)

1. VARIOUS METHODS

Several numerical methods based on finite-difference replacements of the diffusion equation have been proposed. They differ in the way they treat the moving boundary and the grid on which numerical values are calculated. In general, the moving boundary will not coincide with a grid line in successive time steps, δt, if we take δt to be constant and predetermined. Douglas and Gallie [10] chose each δt iteratively so that the boundary always moved from one grid line to the neighbouring one in an interval δt. This means that successive time intervals are of different durations. Crank [4] proposed two methods: the first fixed the boundary by a change of the space variable; the second used special finite-difference formulae based on Lagrangian interpolation for unequal intervals in the neighbourhood of the moving boundary, in order to track its progress between grid lines. Ehrlich [11] employed Taylor expansions in time and space near the boundary. Murray and Landis [16] deformed the grid so that the number of space intervals between the moving boundary and an outer surface remained constant, with a suitable transformation of the diffusion equation. Crank and Gupta [6] and [7] developed several procedures in which the grid is moved with the velocity of the boundary. A recent method [1], [2] and [9] of handling the diffusion equation which offers a special advantage in Stefan problems interchanges the concentration and space variables so that x becomes the dependent and C and t the independent variables. If the boundary is known to occur at a fixed concentration its position comes naturally from the solution without special treatment.

2. A DIFFUSION EXAMPLE

A typical moving boundary problem in diffusion can be formulated as follows. Suppose that an infinite sheet of uniform material of thickness $2a$ is placed symmetrically in a well-stirred solution of extent 2ℓ, and the solute diffuses into the sheet which is initially free of solute. The sheet contains a constant number of sites S per unit volume on each of which one diffusing solute molecule can be instantaneously and permanently captured. The concentration in the solution is always uniform and is initially C_B expressed as the number of molecules per unit volume of solution. The concentration at any time just within the surface of the sheet is taken to be that in the solution. We denote by $C(x,t)$ the concentration in molecules per unit volume of freely-diffusing molecules at time t at a point in the sheet at distance x from the boundary between the sheet and the solution. The concentration of immobilised molecules is zero if C is zero and equal to S when C is non-zero. If we assume the diffusion coefficient D to be constant, the one-dimensional equation

$$\frac{\partial C}{\partial t} = D\frac{\partial^2 C}{\partial x^2} \tag{2.1}$$

is to be solved in $0 < x < X(t)$ subject to

$$X(0) = 0, \tag{2.2}$$

$$C = 0, \quad x = X, \quad t \geqslant 0, \tag{2.3}$$

$$S dX/dt = -(D\partial C/\partial x), \quad x = X, \ t > 0. \tag{2.4}$$

Here $X(t)$ is the value of x at which $C = 0$ and denotes the position of the moving boundary and $C \equiv 0$ in $X \leqslant x \leqslant a$.

Denote by $C_1(t)$ the value of C at the surface and throughout the well-stirred solution at time t. Thus $C_1(t) = C_B$ when $t = 0$ and decreases as diffusion proceeds. The condition of overall conservation of solute is then

$$C_1(t) = C_B - \frac{1}{\ell}\int_0^X (C+S)\,dx, \quad t \geqslant 0. \tag{2.5}$$

3. FIXING THE BOUNDARY

Crank [4] obtained numerical solutions for this problem by using two transformations (*i*) to remove the singularity at $x = 0$, $t = 0$, (*ii*) to fix the advancing boundary.

The singularity is handled by use of the new variables

$$c = C/C_B, \quad s = S/C_B \quad \xi = x/(Dt)^{\frac{1}{2}}, \quad \tau = (Dt/a^2)^{\frac{1}{2}} \tag{3.1}$$

when the relevant equations are

$$2\frac{\partial^2 c}{\partial \xi^2} = \tau\frac{\partial c}{\partial \tau} - \xi\frac{\partial c}{\partial \xi} \tag{3.2}$$

$$c = 0, \quad \xi = \infty, \quad \tau = 0 \tag{3.3}$$

$$-\tfrac{1}{2}\frac{d\xi_X}{d\tau} = \frac{1}{s\tau}\left(\frac{\partial c}{\partial \xi}\right)_{\xi_X} + \frac{\xi_X}{2\tau}, \quad \tau > 0, \tag{3.4}$$

where $\xi_X = X/(Dt)^{\frac{1}{2}}$ is the position of the moving boundary in the transformed variables. Also

$$c = 0, \quad \xi \geqslant \xi_X, \quad \tau \geqslant 0, \tag{3.5}$$

while (2.5) becomes

$$c = 1 - \frac{a}{\ell}\int_0^{\xi_X}(c+s)\tau d\xi, \quad \xi = 0, \quad \tau \geqslant 0. \tag{3.6}$$

When $a \to \infty$, these equations have an analytical solution for $0 \leqslant \xi \leqslant \xi_X$,

$$c = 1 + B\,\mathrm{erf}(\tfrac{1}{2}\xi)$$

which satisfies $c = 1$, $\xi = 0$. The constants B and ξ_X are determined from (3.4) and (3.5):

$$1 + B\,\mathrm{erf}(\tfrac{1}{2}\xi_X) = 0, \tag{3.7}$$

and

$$\pi^{\frac{1}{2}}\xi_X \mathrm{erf}(\tfrac{1}{2}\xi_X)\exp(\tfrac{1}{4}\xi_X^2) = 2/s. \tag{3.8}$$

We now fix the moving boundary by writing

$$\eta = \xi/\xi_X, \quad \tau = \tau. \tag{3.9}$$

Essentially the same transformation was used previously by

Landau [14]* in a consideration of ablating slabs in which the melt is continuously and immediately removed from the surface. Citron [3] extended the treatment to include temperature-dependent thermal properties. Sanders [19] obtained an analytical solution for a particular specified motion of the free boundary. Using (3.9) in the present problem we obtain

$$2\,\frac{\partial^2 c}{\partial\eta^2} = \xi_X^2\,\tau\,\frac{\partial c}{\partial\tau} + \frac{2}{s}\left(\frac{\partial c}{\partial\eta}\right)_1\frac{\partial c}{\partial\eta},\quad 0 < \eta < 1, \tag{3.10}$$

$$-\tfrac{1}{2}\,\frac{d\xi_X}{d\tau} = \frac{1}{s\tau\xi_X}\left(\frac{\partial c}{\partial\eta}\right)_1 + \frac{\xi_X}{2\tau},\quad \eta = 1, \tag{3.11}$$

$$c = 1 - \frac{a}{\ell}\int_0^1 (c+s)\tau\,\dot{\xi}_X\,d\eta,\quad \eta = 0, \tag{3.12}$$

where we have used $(\partial c/\partial\eta)_1$ to mean $(\partial c/\partial\eta)$ at $\eta = 1$. The problem is now that of finding values of ξ_X and $(\partial c/\partial\eta)_1$ which are mutually consistent with (3.11) and for which the solution of (3.10) satisfies (3.12). Crank [4] and [5] used a convenient iterative finite-difference scheme to obtain numerical solutions for a plane sheet, cylinder and sphere in which the fraction $s = S/C_B$ *i.e.* the ratio of the number of sites per unit volume to the number of molecules in unit volume of solution, takes different chosen values. Meadley [15] considered the back diffusion of solute in a layer of solution when the free surface recedes due to evaporation of the solvent. He fixed the receding surface by a transformation essentially the same as (3.9).

4. LAGRANGE INTERPOLATION

We return to the (x,t) plane and develop finite-difference approximations to derivatives based on functional values which are not necessarily equally spaced in the argument.

The Lagrangian interpolation formula is

$$f(x) = \sum_{j=0}^{n} \ell_j(x)\, f(a_j) \tag{4.1}$$

where

*See also Ferriss (p.252).

$$\ell_j(x) = \frac{p_n(x)}{(x-a_j)p'_n(a_j)} , \qquad (4.2)$$

$$p_n(x) = (x-a_0)(x-a_1)\dots(x-a_{n-1})(x-a_n), \qquad (4.3)$$

and $p'_n(a_j)$ is its derivative with respect to x at $x = a_j$. We restrict attention to three-point formulae *i.e.* $n = 2$ and find that

$$\frac{1}{2}\frac{d^2f(x)}{dx^2} = \frac{f(a_0)}{(a_0-a_1)(a_0-a_2)} + \frac{f(a_1)}{(a_1-a_0)(a_1-a_2)} + \frac{f(a_2)}{(a_2-a_0)(a_2-a_1)} \qquad (4.4)$$

and

$$\frac{df(x)}{dx} = \ell_0^1(x)f(a_0) + \ell_1^1(x)f(a_1) + \ell_2^1(x)f(a_2), \qquad (4.5)$$

where

$$\ell_0^1(x) = \frac{(x-a_1)+(x-a_2)}{(a_0-a_1)(a_0-a_2)}, \quad \ell_1^1(x) = \frac{(x-a_2)+(x-a_0)}{(a_1-a_0)(a_1-a_2)}, \qquad (4.6)$$

$$\ell_2^1(x) = \frac{(x-a_0)+(x-a_1)}{(a_2-a_0)(a_2-a_1)} .$$

We apply these formulae in the neighbourhood of the moving boundary. Taking the plane sheet as an example, we consider it to comprise M layers each of thickness δx, and suppose the boundary, $x = X$, at time t to be somewhere in the $(m+1)$th layer (Fig. 1). Thus $X = (m+p)\delta x$ where $0 < p < 1$.

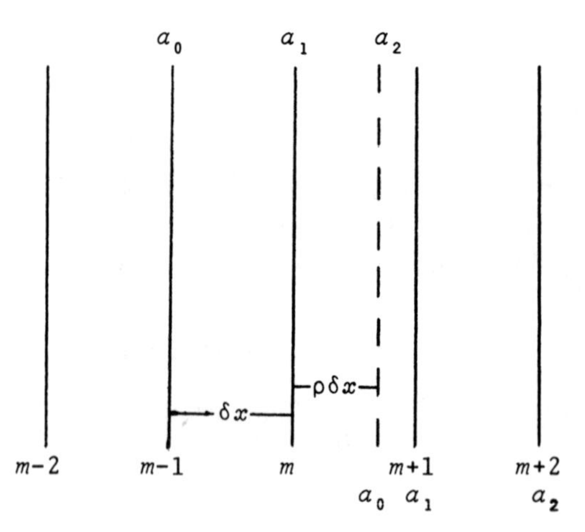

Fig. 1.

If for $x < X$ we identify

$$f(a_0) = c_{m-1}, \quad f(a_1) = c_m, \quad f(a_2) = c_x \qquad (4.7)$$

then (4.4) and (4.5) become

$$\frac{\partial^2 c}{\partial x^2} = \frac{2}{(\delta x)^2}\left\{\frac{c_{m-1}}{p+1} - \frac{c_m}{p} + \frac{c_x}{p(p+1)}\right\}, \quad x = m\delta x, \qquad (4.8)$$

and

$$\frac{\partial c}{\partial x} = \frac{1}{\delta x}\left\{\frac{pc_{m-1}}{p+1} - \frac{(p+1)c_m}{p} + \frac{2p+1}{p(p+1)}c_x\right\}, \quad x = X. \qquad (4.9)$$

Similarly for $x > X$ we have

$$\frac{\partial^2 c}{\partial x^2} = \frac{2}{(\delta x)^2}\left\{\frac{c_x}{(1-p)(2-p)} - \frac{c_{m+1}}{1-p} + \frac{c_{m+2}}{2-p}\right\}, \quad x = (m+1)\delta x, (4.10)$$

and

$$\frac{\partial c}{\partial x} = \frac{1}{\delta x}\left\{\frac{2p-3}{(1-p)(2-p)}c_x + \frac{2-p}{1-p}c_{m+1} - \frac{1-p}{2-p}c_{m+2}\right\}, \quad x = X. \qquad (4.11)$$

We use these formulae for the space derivatives together with the usual explicit or implicit replacements of time derivatives in the diffusion equation itself and in the conditions on the moving boundary, $x = X$. For points other than $m\delta x, X$ and $(m+1)\delta x$, we use the usual finite-difference formula for equal intervals including any condition on $x = 0$.

In the problem posed in §2, $c \equiv 0$ for $x > X$ so that the replacements (4.10) and (4.11) for $x > X$ are not required.

5. A MEDICAL EXAMPLE*

In this problem, oxygen is allowed to diffuse into a medium and some of the oxygen is absorbed and thereby removed from the diffusion process. The oxygen concentration at the surface of the medium is maintained constant. This first phase of the problem continues until a steady state is reached in which the oxygen does not penetrate any further into the medium. The surface is then sealed so that no further oxygen passes in or out. The medium continues to absorb the available oxygen already in it and consequently the boundary marking the furthest depth of penetration recedes towards the sealed surface.

*See Crank (p.65).

The diffusion-with-absorption process is represented by the equation

$$\frac{\partial C}{\partial T} = D \frac{\partial^2 C}{\partial X^2} - m, \tag{5.1}$$

where $C(X,T)$ denotes the concentration of the oxygen free to diffuse at a distance X from the outer surface at time T, D is the diffusion constant and m, the rate of consumption of oxygen per unit volume of the medium, is also assumed constant. The problem has two parts.

(*i*) Steady state

During the initial phase, when oxygen is entering through the surface, the boundary condition there is

$$C = C_0, \quad X = 0, \quad T \geqslant 0, \tag{5.2}$$

where C_0 is constant.

A steady-state is ultimately achieved in which $\partial C/\partial T = 0$ everywhere when both the concentration and its space derivative are zero at a point $X = X_0$. No oxygen diffuses beyond this point and we have

$$C = \partial C/\partial X = 0, \; X \geqslant X_0. \tag{5.3}$$

The required solution in the steady-state is easily found to be

$$C = \frac{m}{2D}(X-X_0)^2 \tag{5.4}$$

where $X_0 = \sqrt{(2DC_0/m)}$.

(*ii*) Moving boundary problem *

After the surface has been sealed the point of zero concentration, originally at $X = X_0$, recedes towards $X = 0$. This second phase of the problem can be expressed in terms of the variables

$$x = \frac{X}{X_0}, \quad t = \frac{DT}{X_0^2}, \quad c = \frac{DT}{mX_0^2} = \frac{C}{2C_0} \tag{5.5}$$

by the equations

$$\frac{\partial c}{\partial t} = \frac{\partial^2 c}{\partial x^2} - 1, \quad 0 \leqslant x \leqslant x_0(t), \tag{5.6}$$

---

*See Ockendon (p.141), Ferriss (p.251).

$$\partial c/\partial x = 0, \quad x = 0, \quad t \geqslant 0, \tag{5.7}$$

$$c = \partial c/\partial x = 0, \quad x = x_0(t), \quad t \geqslant 0, \tag{5.8}$$

$$c = \tfrac{1}{2}(1 - x)^2, \quad 0 \leqslant x \leqslant 1, \quad t = 0. \tag{5.9}$$

The surface is sealed at $t = 0$ and $x_0(t)$ denotes the position of the moving boundary in the reduced space variable. An approximate analytical solution can be used to get away from the singularity at $x = 0$, $t = 0$ due to the instantaneous sealing of the surface [6]. In the range $0 \leqslant t \leqslant 0.020$ the expression

$$c(x,t) = \tfrac{1}{2}(1-x)^2 - 2\sqrt{\left(\frac{t}{\pi}\right)}\exp\left\{-\left(\frac{x}{2\sqrt{t}}\right)^2\right\} + 2\operatorname{erfc}\left(\frac{x}{2\sqrt{t}}\right) \tag{5.10}$$

$$0 \leqslant x \leqslant 1,$$

is sufficiently accurate for most purposes. During this time the boundary has not moved to within the accuracy of the calculations. To follow its subsequent movement Crank and Gupta [6] first used Lagrangian interpolation and a Taylor series near the moving boundary.

6. A MOVING GRID SYSTEM

In a second paper Crank and Gupta [7] used a grid system which moved with the velocity of the moving boundary. This had the effect of transferring the unequal space intervals from the neighbourhood of the moving boundary to the sealed surface. An improved degree of smoothness in the calculated motion of the boundary was obtained compared with the results of using Lagrange interpolation near the moving boundary.

We divide the region $0 < x < 1$ into n intervals each of width Δx such that $x_i = i\Delta x$, $i = 0,1,\ldots,n$ and $n\Delta x = 1$. We denote by c_i^j the values of c at $(i\Delta x, j\Delta t)$, $j = 0,1,2,\ldots$. We start at $t = 0$ with the small-time solution (5.10).

In the first time interval Δt we evaluate c_{n-1}^1 in the usual way by the 3-point explicit, finite-difference formula for evenly spaced data. Crank and Gupta [6] showed that a Taylor expansion about the moving boundary leads to a simple relationship for its movement, ε_1, in the first interval given by

$$\varepsilon_1 = \Delta x - \sqrt{(2c_{n-1}^1)}. \tag{6.1}$$

The whole grid is now moved a distance ε_1 to the left as in Fig. 2.

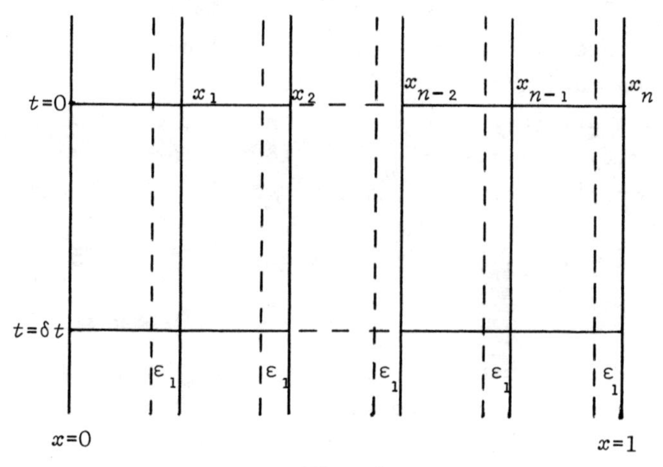

Fig. 2.

We now wish to interpolate the values of c_i^0 and the second space derivatives at each of the points $x_1-\varepsilon_1$, $x_2-\varepsilon_2,\ldots,x_{n-1}-\varepsilon_1$, $1-\varepsilon_1$ at $t = 0$. Crank and Gupta [7] describe two methods of doing this, one based on cubic splines which involves the solution of a tridiagonal set of algebraic equations, and a simpler method using a cubic approximation in each space interval fitted to values of c and its second space derivatives at each grid point.

Having carried out the interpolation we can proceed to time $2\Delta t$ by working on the displaced grid that has uniform spacings except for the one adjacent to the surface $x = 0$ and for which we use Lagrangian formulae.

It is likely, though no rigorous analysis has been tried, that the improved smoothness in the results obtained by moving the whole grid system is a consequence of the different boundary conditions on the sealed surface and on the moving boundary.

7. MURRAY-LANDIS METHOD

Murray and Landis [16] deformed the grid in the space

direction by compressing or stretching it. They kept the number of space intervals between $x = 0$ and $x = x_0$ constant and equal to n, say. Thus $\delta x = x_0(t)/n$ is different in each time step. They differentiated partially with respect to time, t, following a given grid point instead of at constant x. Thus we have for the point $i\delta x$

$$\left(\frac{\partial c}{\partial t}\right)_i = \left(\frac{\partial c}{\partial x}\right)_t \left(\frac{dx}{dt}\right)_i + \left(\frac{\partial c}{\partial t}\right)_x . \tag{7.1}$$

Murray and Landis assumed a general grid point at x to move with velocity dx/dt where

$$\frac{1}{x}\frac{dx}{dt} = \frac{1}{x_0}\frac{dx_0}{dt} . \tag{7.2}$$

The equation (5.6) for example then becomes

$$\left(\frac{\partial c}{\partial t}\right)_i = \frac{x_i}{x_0}\frac{dx_0}{dt}\frac{\partial c}{\partial x} + \frac{\partial^2 c}{\partial x^2} - 1. \tag{7.3}$$

Finite-difference solutions can be obtained when dx_0/dt is determined by an appropriate boundary condition.

Gupta [12] used a Taylor expansion in both x and t to obtain function values in successive time steps at points on a grid system which has the velocity of the moving boundary. His equation is effectively the finite-difference equivalent of (7.3) with $x_i/x_0 = 1$. He compared his results with those obtained by the Murray-Landis method for the oxygen absorption problem described in §5.

8. ISOTHERM MIGRATION METHOD

The aim of all the methods described so far has been to determine how the concentration at a given point changes with time. An alternative method was proposed by Chernous'ko [1] and [2] and independently by Dix and Cizek [9]. It has a particular advantage where phase changes are involved, though its use is not confined to moving boundary problems. In this method the way in which a fixed temperature moves through the medium is calculated. Hence it is referred to as the "isotherm migration method". In the diffusion analogy we trace the motion of contours of constant concentration.

When a phase change or other boundary occurs at a prescribed temperature its motion emerges naturally from the solution without any special treatment. Also the need to evaluate temperature or concentration dependent parameters at each time step is avoided. The appropriate values are simply carried along with the isotherm. These advantages are partly offset by some increase in complexity of the finite-difference equations and difficulties with some boundary conditions.

Instead of expressing the concentration c as a function of x and t, we now re-write the equations so that x becomes a function of c and t. The dependent variable becomes $x(c,t)$ instead of $c(x,t)$. We note that with reference to the diffusion equation in one dimension we can write

$$\left(\frac{\partial c}{\partial x}\right)_t = \left(\frac{\partial x}{\partial c}\right)_t^{-1} \text{ and } \left(\frac{\partial x}{\partial t}\right)_c = -\left(\frac{\partial c}{\partial t}\right)_x \left(\frac{\partial x}{\partial c}\right)_t .$$

Starting from the usual diffusion equation

$$\frac{\partial c}{\partial t} = D\,\frac{\partial^2 c}{\partial x^2} \tag{8.1}$$

and using the above relations we find

$$\frac{\partial x}{\partial t} = D\left(\frac{\partial x}{\partial c}\right)^{-2} \frac{\partial^2 x}{\partial c^2} . \tag{8.2}$$

If we consider as an example, the simplest form of the problem discussed above in §2 in which the condition on $x = 0$ is replaced by

$$c = c_0 = \text{constant}, \quad x = 0, \quad t \geqslant 0, \tag{8.3}$$

then we can re-write the condition (2.4) on the moving boundary as

$$S\,\frac{dX}{dt} = -D\left(\frac{\partial x}{\partial c}\right)^{-1}, \quad c = 0, \quad t \geqslant 0, \tag{8.4}$$

remembering that we took $c = 0$ on $x = X(t)$.

The condition (8.3) becomes

$$x = 0, \quad c = c_0, \quad t \geqslant 0. \tag{8.5}$$

In order to start the solution we need to use an analytic or approximate solution suitable for small times. This provides values of x for equally spaced values of c at a given small time t from which we can proceed in steps, δt, to obtain values of x on a grid in the c,t plane by applying the usual finite difference formulae to the transformed equations.

If we let x_i^n be the value of x at $c = i\delta c$, $t = n\delta t$, we find the diffusion equation (8.2) becomes approximately

$$x_i^{n+1} = x_i^n + 4\delta t\left\{\frac{x_{i+1}^n - 2x_i^n + x_{i-1}^n}{(x_{i-1}^n - x_{i+1}^n)^2}\right\} \tag{8.6}$$

and from (8.4) we have

$$X^{n+1} = X^n - \frac{D\delta t\delta c}{S(x_0^n - x_1^n)} \,. \tag{8.7}$$

Crank and Phahle [8] obtained results by this method for the problem of melting a block of ice that agreed well with the exact analytical solution and with solutions obtained by Goodman's integral method and Lagrangian interpolation.

9. ITERATIVELY VARIED TIME STEP

Finally, we refer to an early method used by Douglas and Gallie [10]. They realised that the moving boundary will not coincide with a grid point if the steps in time are constant and predetermined in length. Instead they adjusted δt iteratively so that each time interval is the time required for the boundary to move through one space interval, δx.* They considered the problem

$$\frac{\partial c}{\partial t} = \frac{\partial^2 c}{\partial x^2}, \quad 0 < x < x_0, \quad t > 0, \tag{9.1}$$

$$\frac{\partial c}{\partial x} = -1, \quad x = 0, \quad t > 0, \tag{9.2}$$

$$u = 0, \quad x \geqslant x_0, \quad t > 0. \tag{9.3}$$

$$\frac{dx_0}{dt} = -\frac{\partial c}{\partial x}, \quad x = x_0, \quad t > 0. \tag{9.4}$$

Thus $\int_{w(x)}^{t} c(x,\tau)\,d\tau$ satisfies a problem similar to (5.6) to

*See also Gelder and Guy (p.78).

(5.9), where $x_0(w(x)) \equiv x$.

They integrated (9.1) to obtain

$$\frac{\partial}{\partial t}\int_0^{x_0} c\ dx = \left[\frac{\partial c}{\partial x}\right]_0^{x_0} = -\frac{dx_0}{dt} + 1$$

which on integrating with respect to t becomes

$$t = x_0 + \int_0^{x_0} c dx. \tag{9.5}$$

Douglas and Gallie used fully implicit formulae, first with an arbitrary value of δt. We denote by δt_n^r, the rth iteration for δt for the time step in which the boundary moves from $n\delta x$ to $(n+1)\delta x$. Then we have a usual set of equations

$$\frac{c_{m,n+1}^r - c_{m,n}}{\delta t_n^r} = \frac{(c_{m-1} - 2c_m + c_{m+1})_{n+1}^r}{(\delta x)^2} \tag{9.6}$$

where $c_{m,n+1}^r$ is the rth estimate of c at $m\delta x$ after $n+1$ time steps. We have the additional conditions

$$c_{n+1,n+1}^r = 0 \tag{9.7}$$

because the boundary is by definition at $(n+1)\delta x$ and

$$c_{0,n+1}^r - c_{1,n+1}^r = \delta x \tag{9.8}$$

is a simple replacement for (9.2).

We solve a set of $n+2$ equations for $n+2$ unknown c^r's with an estimated value δt_n^r, which is then improved iteratively using a simple discrete form of (9.5) namely

$$\delta t_n^{r+1} = (n+1 + c_{0,n+1}^r + \sum_{m=1}^{n} c_{m,n+1}^r)\delta x - t_n. \tag{9.9}$$

Douglas and Gallie started the solution with the moving boundary at $x = \delta x$ by using

$$c_{0,0} = 0,\quad c_{1,1} = 0,\quad c_{0,1} = \delta x,\quad \delta t_0 = \delta x$$

from (9.2) and (9.4).

10. A TWO-DIMENSIONAL PROBLEM

One way of extending the IMM to two space dimensions is promising. Consider the problem

$$\frac{\partial c}{\partial t} = \frac{\partial^2 c}{\partial x^2} + \frac{\partial^2 c}{\partial y^2}, \tag{10.1}$$

$$c = 0 \text{ on } g(x,y) \equiv (x^2 - 1)(y^2 - 1) = 0, \quad t \geqslant 0 \tag{10.2}$$

$$c = 1 \text{ on } f(x,y,t) = 0, \quad t > 0 \tag{10.3}$$

where $f(x,y,t)$ is the contour of the moving interface at time t. We are concerned with a square region lying between $-1 \leqslant x \leqslant 1$, $-1 \leqslant y \leqslant 1$ throughout which $c = 1$ at $t = 0$ which implies

$$f(x,y,0) = 0. \tag{10.4}$$

On the moving surface we have

$$\frac{\partial c}{\partial n} = -\beta V_n, \text{ on } f(x,y,t) = 0, \tag{10.5}$$

where n is the normal to $f(x,y,t) = 0$, V_n is the velocity of the interface in the direction of n and β is a physical constant. By combining (10.1) with

$$\left(\frac{\partial y}{\partial t}\right)_{c,x} = \left(\frac{\partial c}{\partial t}\right)_{x,y} \left(\frac{\partial y}{\partial x}\right)_{t,c} \left(\frac{\partial x}{\partial c}\right)_{y,t} \tag{10.6}$$

and

$$\frac{\partial^2 c}{\partial y^2} = -\frac{\partial^2 y}{\partial c^2}\left(\frac{\partial y}{\partial c}\right)^{-3}, \quad \left(\frac{\partial y}{\partial c}\right)_{x,t} = -\left(\frac{\partial y}{\partial x}\right)_{t,c}\left(\frac{\partial x}{\partial c}\right)_{y,t} \tag{10.7}$$

we obtain

$$\left(\frac{\partial y}{\partial t}\right)_{c,x} = \left\{-\frac{\partial^2 c}{\partial x^2} + \frac{\partial^2 y}{\partial c^2}\left(\frac{\partial y}{\partial c}\right)^{-3}\right\}\left(\frac{\partial y}{\partial c}\right). \tag{10.8}$$

Essentially, this equation expresses the movement of the isotherms in the y direction with x held constant. We compute values of y in successive time intervals, δt, at points on a grid with spacings δc, δx in the (c,x) plane.

In order to obtain $\partial y/\partial t$ on the interface we follow Patel [17] and differentiate (10.3) with respect to t to obtain

$$\frac{\partial c/\partial t}{\partial f/\partial t} = \frac{\partial c/\partial x}{\partial f/\partial x} = \frac{\partial c/\partial y}{\partial f/\partial y}. \tag{10.9}$$

Remembering that

$$V_n = \frac{-\ \partial f/\partial t}{\left[\left(\frac{\partial f}{\partial x}\right)^2 + \left(\frac{\partial f}{\partial y}\right)^2\right]^{\frac{1}{2}}} \tag{10.10}$$

and

$$\frac{\partial c}{\partial n} = \frac{\frac{\partial f}{\partial x}\cdot\frac{\partial c}{\partial x} + \frac{\partial f}{\partial y}\frac{\partial c}{\partial y}}{\left[\left(\frac{\partial f}{\partial x}\right)^2 + \left(\frac{\partial f}{\partial y}\right)^2\right]^{\frac{1}{2}}}, \tag{10.11}$$

and using (10.5) we obtain from (10.9)

$$\frac{\partial c}{\partial t} = +\frac{1}{\beta}\left\{\left(\frac{\partial c}{\partial x}\right)^2 + \left(\frac{\partial c}{\partial y}\right)^2\right\}. \tag{10.12}$$

By using (10.6) and the second of (10.7) together with (10.12) we find

$$\frac{\partial y}{\partial t} = -\frac{1}{\beta}\left\{1 + \left(\frac{\partial y}{\partial x}\right)^2\right\}\left(\frac{\partial y}{\partial c}\right)^{-1} \tag{10.13}$$

at the interface $f(x,y,t) = 0$.

In principle we use (10.8) at general points of the grid but (10.13) at the interface. Computational details are in process of publication. A convenient method of starting the IMM numerical solution is provided by an integral method based on a hyperbolic contour for the interface [13] or on the two-parameter contour used by Poots [18].

REFERENCES

1. Chernous'ko, F.L., *Zh. Prikl. Mekh. i. Tekhn. Fiz.* No. 2, 6 (1969).
2. Chernous'ko, F.L., *Int. Chem. Eng.* **10**, No. 1, 42 (1970).
3. Citron, S.J., *J. Aero/Space Sci.* **27**, 219, 317, 470 (1960).
4. Crank, J., *Quart. J. Mech. Appl. Maths.* **10**, 220 (1957).
5. Crank, J., *Trans. Faraday Soc.* **53**, 1083 (1957).
6. Crank, J. and Gupta, R.S., *J. Inst. Maths. Applics.* **10**, 19 (1972).

7. Crank, J. and Gupta, R.S., *J. Inst. Maths. Applics.* **10**, 296 (1972).

8. Crank, J. and Phahle, R.D., *Bull. IMA.*, **9**, 12 (1973).

9. Dix, R.C. and Cizek, J., in "Heat Transfer 1970" **1**, pCu.11 (Grigull, U. and Hahne, E. editors) Elsevier (1970).

10. Douglas, J. and Gallie, T.M. *Duke Math. J.* **22**, 557 (1955).

11. Ehrlich, L.W., *J. Assoc. Comp. Machinery*, **5**, 161 (1958).

12. Gupta, R.S., "Computer Methods in Appl. Mechs. and Eng.," (submitted for publication) (1974).

13. Gupta, R.S., Brunel University Technical Report TR/40 (1974).

14. Landau, H.G., *Quart. Appl. Maths.* **8**, 81 (1950).

15. Meadley, C.K., *Quart. J. Mech. and Appl. Math.* **24**, 43 (1971).

16. Murray, W.D. and Landis, F., *J. Heat Transfer* **81**, 106 (1959).

17. Patel, P.D., *AIAA Journal* **6**, 2454 (1968).

18. Poots, G., *Int. J. Heat Mass Transfer* **5**, 339 (1962).

19. Sanders, R.W., *Am. Rocket Soc. J.* **30**, 1030 (1960).

Heat Balance Methods in Melting Problems

B. Noble
*(University of Oxford)**

Heat balance methods exploit the principle that the rate of increase of the amount of heat in a given volume is equal to the rate of production of heat in the volume minus the rate at which heat is lost across the boundary [1]. By guessing suitable forms for the distribution of temperature in space, a system of ordinary differential equations is obtained for the functions of time that were left as arbitrary coefficients in the original guess. In one-dimensional problems, the spatial dependence is usually taken to be quadratic on either side of the phase boundary.

In this presentation, heat balance methods were viewed as a method for approximate solution of the partial differential equation formulation of the problem. For simple one-dimensional examples, excellent agreement was obtained with other numerical methods. Various ways of extending the range of applicability of the method were described. Rather than introducing cubic or higher order spline functions of position, the most promising way of systematically improving the accuracy of the approximation was shown to be by repeated spatial subdivision with the use of quadratic approximations in each interval. Such a "finite element" approach could play a key rôle in the use of heat balance methods in more than one space dimension.

There was still a need for a detailed evaluation of improved heat balance methods with conventional finite difference approaches, both in one or more space dimensions

---

*Present address: Mathematics Research Center, University of Wisconsin.

and especially in situations involving nonlinear thermal conductivities. The accuracy of the basic method has been assessed in [2].

REFERENCES

1. Goodman, T.R., "Application of Integral Methods to Transient Nonlinear Heat Transfer", Advances in Heat Transfer, Vol. 1., 51-122, Academic Press (1964).
2. Langford, D., "The Heat Balance Integral Method", *Int. J. Heat Mass. Tranfer.*, 16, 2424-2428 (1973).

WHAT ARE THE BEST NUMERICAL METHODS?

L. Fox
(University of Oxford)

1. INTRODUCTION

There are very few areas of numerical analysis in which the question of my title can be answered with authority and conviction. Problems of the same general nature can differ enough in detail to make a good method for one problem less satisfactory and even mediocre for another almost similar problem. Even in linear algebra, in which the number of different problems is respectably small, we have quite a lot of different methods associated with different particular problems. In ordinary differential equations of initial-value type, moreover, which is a field far less complicated than the one we are discussing here, it is only in very recent times that we have begun a systematic evaluation of methods, and generally this is being done in the rather pedestrian way of comparing the relative performance of methods on selected test examples, as far as possible under the same conditions of computer hardware and software.

Numerical analysts, moreover, tend to think of the evaluation of numerical methods in respect of the time taken to get six accurate figures, and whether or not one can provide a computable error bound. In many practical cases we do not need six figures, and even an order of magnitude may be all that is required. A good method for six figures need not be a good method for an order of magnitude, and experience may indicate the value of methods for which error bounds are unattainable in our present state of knowledge.

A final introductory remark is that I do not believe that my title is the best possible. For in the first place the best numerical methods may be those which associate neatly and in relevant contexts with the best analytical methods, and in the second place approximate analytical methods may be quite superior in some contexts to <u>any</u> numerical method. I am trying to say that what we should be seeking are the best methods, which will sometimes be analytical, sometimes numerical, and most often a combination of the two.

How then are we to achieve this? I make no apologies for not knowing the answer, because when we were discussing this topic several months ago we estimated that the amount of reading and computation necessary to produce any worthwhile conclusions was an order of magnitude greater than the time available to the date of this conference. All I have been able to do is to read some of the literature and get some ideas on what I think are the more important developments. From here I think we should build on the material of this conference by trying to assemble the many and various problems into a relatively small number of models, then to assemble the various possible methods, and then to subject them to practical tests in the way that various workers are now trying to do for ordinary differential equations.

With this introduction I would like to consider the various possible methods, concentrating for convenience on the Stefan problem in heat transfer. Most of these methods have been mentioned by other speakers, but I am sure that it will do no harm to mention them again, usually by means of simple examples, as an aid to some sort of relative evaluation.

2. FINITE-DIFFERENCE METHODS FOR THE STEFAN PROBLEM

2.1 Basic methods in one or more space dimensions

In one space dimension, the usual Stefan problem, with fairly obvious notation, is represented typically by the differential equations

$$\left.\begin{aligned}\frac{\partial}{\partial x}\left(k_1\frac{\partial u_1}{\partial x}\right) &= \rho_1 c_1\,\frac{\partial u_1}{\partial t},\quad 0 \leqslant x \leqslant s(t)\\ \frac{\partial}{\partial x}\left(k_2\frac{\partial u_2}{\partial x}\right) &= \rho_2 c_2\,\frac{\partial u_2}{\partial t},\quad x \geqslant s(t)\end{aligned}\right\},\tag{2.1}$$

with conditions at the interface $x = s(t)$ of the type

$$u_1 = u_2 = u_m \tag{2.2}$$

$$k_2\frac{\partial u_2}{\partial x} - k_1\frac{\partial u_1}{\partial x} = -\rho_1 L\frac{ds(t)}{dt}\ . \tag{2.3}$$

Here 1 refers to solid and 2 to liquid (Fig. 1). There will also be some boundary conditions at $x = 0$ and at $x = \ell$ or as $x \to \infty$, and an initial condition at $t = 0$. Our problem is the determination of $u_1(x,t)$, $u_2(x,t)$ and $s(t)$.

Now there is no particular difficulty in writing down and solving suitable finite-difference equations, with intervals Δx and Δt, in the main parts of the liquid and solid regions. The problems arise near the interface, partly due to the discontinuity in the temperature gradient at the interface, and partly due to the fact that the x-coordinate of a new mesh point on the line $t + \Delta t$ is not known in advance. It is therefore not known *a priori* which region is inhabited by points like P and Q in Fig. 1. This information is implicit in the differential equations and interface conditions, and will usually require some iterative process for the solution of the nonlinear system.

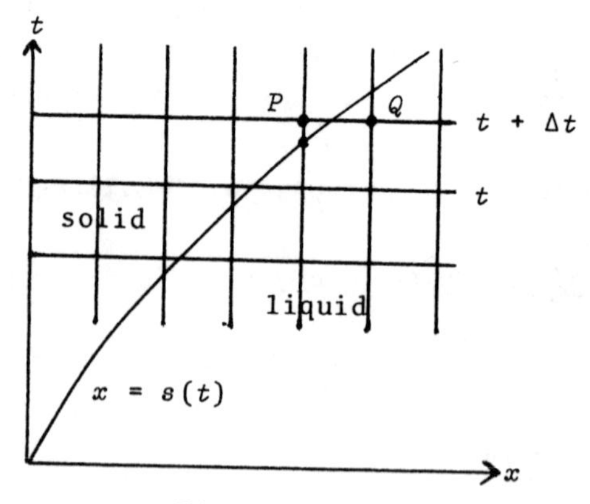

Fig. 1.

We also require, of course, that our finite-difference equations be everywhere reasonably accurate, that is that the local truncation error is small, and also that our step-by-step process in the t-direction is stable, so that the accumulation of local errors is not excessive and so that, as Δt, $\Delta x \to 0$, our computed solution tends to the correct solution everywhere.

The complexity of the finite-difference equations becomes apparent when we look at some published methods. For example Douglas and Gallie [11]* considered the relatively easy problem defined by the equations

$$\frac{\partial^2 u}{\partial x^2} = \frac{\partial u}{\partial t}, \quad 0 \leqslant x \leqslant s(t), \tag{2.4}$$

$$\frac{\partial u}{\partial x}(0,t) = -1, \quad t > 0, \tag{2.5}$$

$$u(s(t),t) = 0, \quad t \geqslant 0, \tag{2.6}$$

$$s(0) = 0, \tag{2.7}$$

$$\frac{d}{dt}s(t) = -\frac{\partial u}{\partial x}(s(t),t), \tag{2.8}$$

the last equation implying and being used in the form

$$s(t) = t - \int_0^{s(t)} u(x,t)\,dx. \tag{2.9}$$

With Δx and Δt specified they use the general implicit formula, which is known to be stable, given by

$$\frac{u_{i,n+1} - u_{i,n}}{\Delta t} = \frac{u_{i+1,n+1} - 2u_{i,n+1} + u_{i-1,n+1}}{(\Delta x)^2}, \quad i=1,2,\ldots N_n-1, \tag{2.10}$$

where $N_n = [s(t_n)/\Delta x]$, and $u_{r,s} = u(r\Delta x, s\Delta t)$. Near the interface this is replaced by the more complicated equation

$$\frac{u_{N_n,n+1} - u_{N_n,n}}{\Delta t} = \frac{2}{\Delta x}\left\{\frac{u_{N_n-1,n+1}}{s(t_{n+1}) - (N_n-1)\Delta x} - \frac{u_{N_n,n+1}}{s(t_{n+1}) - N_n\Delta x}\right\}, \tag{2.11}$$

which assumes that $u(x)$ behaves like a quadratic near $x = s(t_{n+1})$ and vanishes at that point. The arrangement of points is shown in Fig. 2.

---

* See also Crank (p. 203).

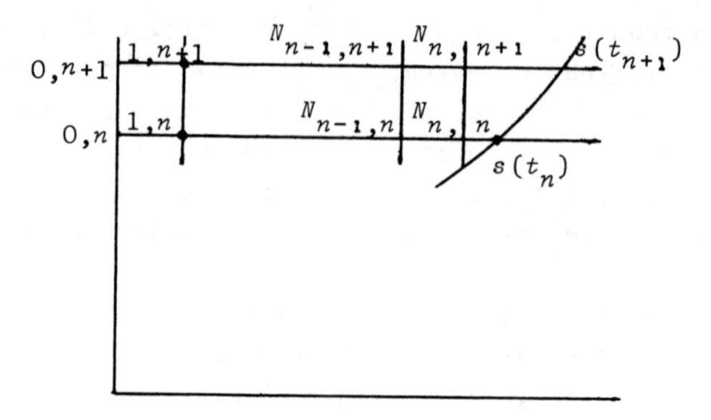

Fig. 2.

There is another relation between $u_{1,n+1}$ and $u_{0,n+1}$, which they take as

$$u_{1,n+1} - u_{0,n+1} = -\Delta x \tag{2.12}$$

to represent the boundary condition, and the other interface condition is represented by a quadrature formula of the type

$$s(t_{n+1}) = t_{n+1} - \sum_{i=1}^{N_n-1} u_{i,n+1}\Delta x - \tfrac{1}{2}\{s(t_{n+1}) - (N_n-1)\Delta x\}u_{N_n,n+1} \tag{2.13}$$

Equations (2.10) to (2.12) are solved simultaneously for the $u_{i,n+1}$, with an estimated $s(t_{n+1})$, and the latter is then improved by substituting the results on the right of (2.13). The paper proves the convergence of the resulting iterative process, and observes that if $s(t_{n+1}) > (N_n + 1)\Delta x$ the "missing" value $u_{N_n+1,n+1}$ is interpolated linearly for use at the next time step.

One can improve on this a bit, for example by using the Crank-Nicolson type implicit method to reduce the truncation error in the t-direction, and using instead of (2.12) the equation

$$\frac{u_{0,n+1} - u_{0,n}}{\Delta t} = \frac{2(u_{1,n+1} - u_{0,n+1} - \Delta x)}{(\Delta x)^2} \tag{2.14}$$

to reduce the truncation error in the x-direction. But the equations are somewhat complicated even for this very simple problem.

For more difficult problems, and in more than one space dimension, the complexities increase by several orders of magnitude. For example, Lazaridis [17], considers a region D_1 of liquid bounded by a fixed surface R_1, and a region D_2 of solid bounded by a fixed surface R_2, the regions being separated by an interface R (Fig. 3).

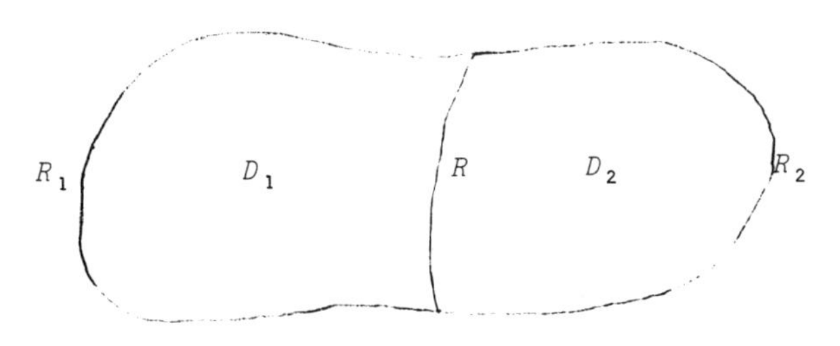

Fig. 3.

His equations (in Cartesian coordinates x_1, x_2 and x_3 and with suitable normalisation) are given by

$$\gamma_i \nabla^2 u_i = \partial u_i/\partial t, \quad i = 1, 2 \tag{2.15}$$

$$S_1(u_1) = 0 \text{ on } R_1, \; S_2(u_2) = 0 \text{ on } R_2, \tag{2.16}$$

with u_1, u_2 and R specified everywhere at $t = 0$. Now $u_1 = u_2 =$ constant on the isothermal surface R. Thus, if R is $x_1 = \varepsilon_1(x_2, x_3, t)$, the heat balance equation can be written as

$$\left\{1 + \left(\frac{\partial\varepsilon_1}{\partial x_2}\right)^2 + \left(\frac{\partial\varepsilon_1}{\partial x_3}\right)^2\right\}\left\{\frac{\partial u_1}{\partial x_1} - K\frac{\partial u_2}{\partial x_1}\right\} = \lambda\frac{\partial\varepsilon_1}{\partial t}, \tag{2.17}$$

for suitable constants K,λ. Equivalent formulae can be obtained by permuting the indices appropriately, and two further equations express the fact that R is an isotherm.

I have neither time nor space to give either these equations or the details of the numerical method. But even in two space dimensions the equations are complicated and cumbersome in the attempt to achieve accuracy near the interface. In Fig. 4, for example, different equations must be used according as Δx_1 and $\Delta x_2 \gtrless \frac{1}{2}$, and according to the location of points like 1, 2, 3, 4, 5 and 6 in relation to the interface. In three dimensions the number of different possibilities is substantially greater, and the reader of the paper is advised to consult Lazaridis

[17] for full details. He claims good results on several examples in two and three dimensions, with computing times measured in a few minutes on the IBM360-75, and if his programs are available for general use they would be worth obtaining.

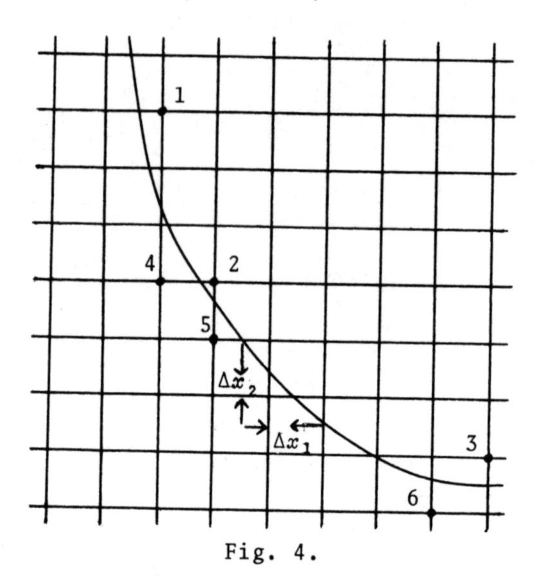

Fig. 4.

I should add that he considers only explicit finite-difference methods (perhaps not surprising in view of the complexity even of these equations), and must therefore use rather small time steps to avoid computational instabilities.

2.2 Modifications

A number of writers have tried to reduce the complications by using special and somewhat *ad hoc* devices near the interface. For example, in the paper by Douglas and Gallie [11] already mentioned, they claim as more computationally practicable a method which fixes Δx but which selects Δt_n at time step n so that $s(t)$ is the right-most mesh point at this stage, as shown in Fig. 5.

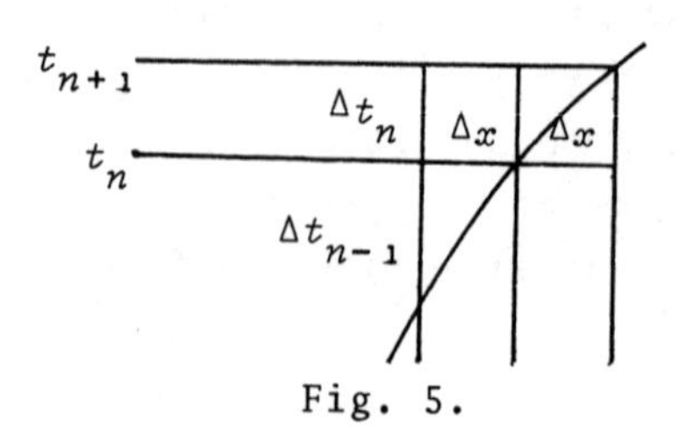

Fig. 5.

Their implicit formulae are now given by

$$\left.\begin{aligned} \frac{u_{i,n+1} - u_{i,n}}{\Delta t_n} &= \frac{u_{i-1,n+1} - 2u_{i,n+1} + u_{i+1,n+1}}{(\Delta x)^2}, \\ u_{0,n+1} - u_{1,n+1} &= \Delta x, \\ u_{n+1,n+1} &= 0 \end{aligned}\right\}, \quad (2.18)$$

with no "special" formulae involved, and the interface condition becomes

$$s(t_{n+1}) = t_{n+1} - \int_0^{s(t_{n+1})} u(x,t_{n+1})\,dx\ . \quad (2.19)$$

Again they use an iterative method with proven convergence, and also remark that iteration can be avoided with judicious use of a combination of (2.19) and the first of (2.18). Though this is clearly an improvement on their other method, (again it could be further improved with formulae of smaller truncation error) we are dealing with an extremely simple problem in one space dimension, and there would seem to be no easy extension to more complicated problems.

Other writers have varied the size of the space intervals. The most recent suggestion is that of Crank and Gupta [8]* who consider the determination of the moving boundary in the problem given by

$$\left.\begin{aligned} \frac{\partial u}{\partial t} &= \frac{\partial^2 u}{\partial x^2} - 1, \ 0 \leqslant x \leqslant s(t), \\ \frac{\partial u}{\partial x} &= 0, \ x = 0, \\ u &= \tfrac{1}{2}(1-x)^2, \ t = 0, \ 0 \leqslant x \leqslant 1, \\ u &= \frac{\partial u}{\partial x} = 0, \ x = s(t) \end{aligned}\right\}. \quad (2.20)$$

The moving boundary is at $x = 1$ at $t = 0$, and at time Δt it has moved a distance ε to the left, as shown in Fig. 6. To avoid complications at this point the whole grid system is moved a distance ε to the left, so that the unequal interval problem is transferred to the more tractable region near $x = 0$. Values like $u(x_{n-1}, \Delta t)$ are computed from explicit finite-difference formulae, and if ε is known we can then interpolate

---

* See also p. 65.

to compute the values at the "dotted" grid points for use at the next step.

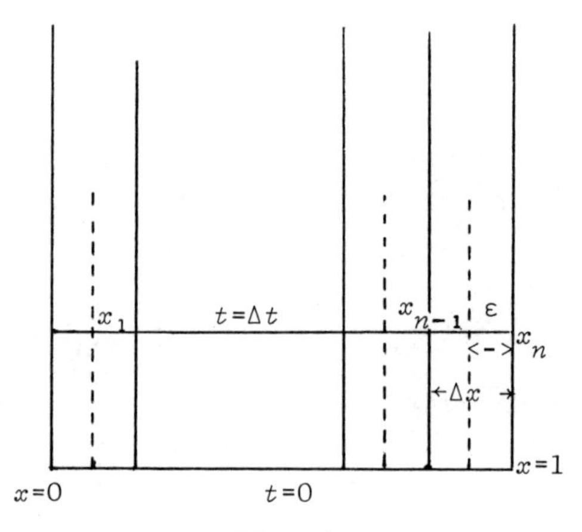

Fig. 6.

The value of ε is obtained from the Taylor series for $u(x_{n-1},\Delta t)$ about the moving point. If $\Delta x - \varepsilon = \ell$, then

$$u(x_{n-1},\Delta t) = (u - \ell \frac{\partial u}{\partial x} + \tfrac{1}{2}\ell^2 \frac{\partial^2 u}{\partial x^2} + \ldots), \qquad (2.21)$$

everything on the right being evaluated at the moving boundary point. In a previous paper Crank and Gupta [7] show that at this point

$$\frac{\partial^2 u}{\partial x^2} = 1, \ \frac{\partial^3 u}{\partial x^3} = -\frac{ds(t)}{dt}, \ \ldots \qquad (2.22)$$

so that

$$u(x_{n-1},\Delta t) = \tfrac{1}{2}\ell^2 + \tfrac{1}{6}\ \ell^3 \frac{dx(t)}{dt} + \ldots\ . \qquad (2.23)$$

If the boundary is not moving too quickly a reasonable approximation is

$$\ell = \sqrt{2u(x_{n-1},\Delta t)}, \qquad (2.24)$$

and the remaining computational details are relatively trivial.

It is not obvious how to extend the method of Crank and Gupta to more complicated situations and to more than one space dimension. The method also has the usual stability problems of explicit techniques, and the corresponding implicit technique could increase by an order of magnitude the complexity of the arithmetic operations.

In this problem, incidentally, the velocity of the moving boundary does not appear explicitly in the interface condition. Such problems are called "implicit free boundary problems" by Sackett [20]. He seems to imply that the explicit presence of a velocity term is an important ingredient of methods* which use integral equations, and without this term such methods are difficult both in theory and in practice. I mention this again in a later section.

He considers the problem defined by the equations

$$\left.\begin{aligned} \frac{\partial^2 u}{\partial x^2} &= \frac{\partial u}{\partial t}, \quad 0 < x < s(t), \\ u(0,t) &= f_1(t), \quad u(x,0) = \psi(x), \\ u &= f_2(t), \quad \frac{\partial u}{\partial x} = g(t) \text{ on } x = s(t); \ s(0) = a, \end{aligned}\right\} \quad (2.25)$$

a picture of which is shown in Fig. 7.

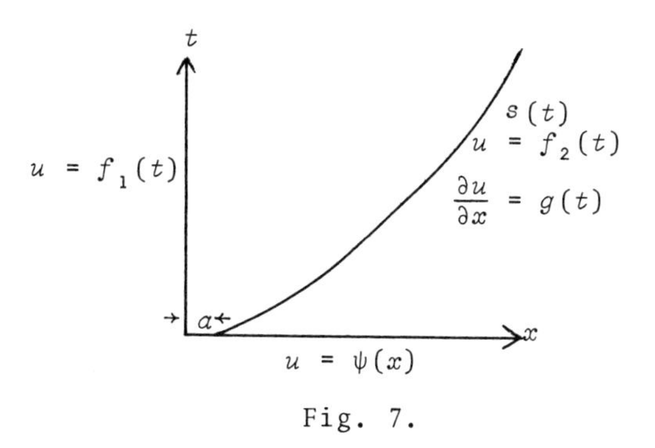

Fig. 7.

His numerical method is defined by the second-order ordinary differential equation

$$\left.\begin{aligned} \frac{d^2 u^{n+1}}{dx^2} &= \frac{1}{\Delta t}\,(u^{n+1} - \bar{u}^n) \\ \bar{u}^n &= u^n, \quad 0 \leqslant x \leqslant s(t_n) \\ \bar{u}^n &= f_2^n + g^n(x - s(t_n)), \text{ if } x > s(t_n) \end{aligned}\right\}, \quad (2.26)$$

* See Tayler (p. 125).

(the latter equation representing a linear extension of u^n for $x > s(t_n)$), together with the three conditions

$$u^{n+1} = f_1^{n+1} \text{ on } x = 0; \; u^{n+1} = f_2^{n+1}, \; \frac{d}{dx} u^{n+1} = g^{n+1} \text{ on } x = s(t_{n+1}). \tag{2.27}$$

He observes that (2.26) has an analytical solution

$$\left.\begin{aligned} u^{n+1}(x) &= A\,\phi_1(x) + B\,\phi_2(x) - \frac{1}{\sqrt{\Delta t}} \int_0^x \bar{u}^n(\xi)\phi_2(x-\xi)d\xi, \\ \phi_1(x) &= \cosh(x/\sqrt{\Delta t}), \; \phi_2(x) = \sinh(x/\sqrt{\Delta t}) \end{aligned}\right\}, \tag{2.28}$$

in which A and B are chosen to satisfy two of (2.27). The third of (2.27) then gives the equation

$$f_2^{n+1}\,\phi_1(X) - \sqrt{\Delta t}\,g^{n+1}\,\phi_2(X) - f_1^{n+1} - \frac{1}{\sqrt{\Delta t}} \int_0^X \bar{u}^n(\xi)\phi_2(\xi)d\xi = 0 \tag{2.29}$$

for the implicit determination of the required $X = s(t_{n+1})$. The required $u^{n+1}(x)$ is then obtained by simple quadrature from (2.28).

Sackett uses this technique to prove that the original problem has a unique solution under certain conditions, and this is his main purpose. As a general numerical technique it has some disadvantages, notably the lack in general of an analytical solution of type (2.28) and the consequent need for discretisation in the x-direction. This then will reintroduce the common difficulties at the free boundary.

2.3 *Important modifications*

(*a*) *Reduction to ordinary 2-point problems*

Sackett's idea of discretising only in the time direction is not of course new, but it has also been exploited by Meyer [18] for the more complicated problem defined by the equations

$$\begin{aligned} &\frac{\partial^2 u}{\partial x^2} + a(t,x)\frac{\partial u}{\partial x} - b(t,x)u - c(t,x)\frac{\partial u}{\partial t} = f(t,x), \\ &u = g(t)\frac{\partial u}{\partial x} + h(t) \text{ on } x = 0, \\ &u = u_0(x) \text{ at } t = 0, \\ &\left.\begin{aligned} &u = 0, \\ &\frac{ds}{dt} + m(x,t)\frac{\partial u}{\partial x} = \ell(x,t) \end{aligned}\right\} \text{ on } x = s(t). \end{aligned} \tag{2.30}$$

He also discretises in the t-direction only, replacing time derivatives by backward difference expressions, and produces the ordinary differential system

$$\left.\begin{aligned} &u_k'' + a_k u_k' - b_k u_k - \frac{c_k}{\Delta t}(u_k - u_{k-1}) = f_k \\ &u_k(0) = g_k u_k'(0) + h_k \\ &u_k(s_k) = 0 \\ &s_k - s_{k-1} + \Delta t\, m_k(s_k)\, u_k'(s_k) = \Delta t\, \ell_k(s_k) \end{aligned}\right\}, \qquad (2.31)$$

where s_k is the x-coordinate of the moving boundary at time $k\Delta t$. This is a two-point boundary-value problem, the extra associated condition being used to compute the position s_k of the boundary.

Meyer now borrows a device which I think was invented some time ago by Hartree and Ridley (though Meyer makes no reference to this work) in which the second-order operator is virtually split into two first-order operators. Writing

$$u_k = W_k v_k + w_k, \quad u_k' = v_k, \qquad (2.32)$$

the problem is reduced to the forward solution of the first-order system

$$\left.\begin{aligned} &W_k' = 1 + a_k W_k - \left(\frac{c_k}{\Delta t} + b_k\right) W_k^2, \quad W_k(0) = g_k \\ &w_k' = -\left(\frac{c_k}{\Delta t} + b_k\right) W_k w_k + \left(\frac{c_k}{\Delta t} u_{k-1} - f_k\right) w_k, \quad w_k(0) = h_k \end{aligned}\right\} \qquad (2.33)$$

from which s_k is obtained by inverse interpolation from the equation

$$s_k - s_{k-1} + \Delta t\, m_k(s_k) v_k(s_k) = \Delta t\, \ell_k(s_k), \qquad (2.34)$$

and the solution is completed by the backward solution of the first-order equation

$$\begin{aligned} &v_k' = \left\{\left(\frac{c_k}{\Delta t} + b_k\right) W_k - a_k\right\} v_k + \left(\frac{c_k}{\Delta t} + b_k\right) w_k + \left(f_k - \frac{c_k}{\Delta t} u_{k-1}\right), \\ &v_k(s_k) = -\frac{w_k(s_k)}{W_k(s_k)} . \end{aligned} \qquad (2.35)$$

This would appear to be a very useful method in one-space dimension, since there are very good programs in existence for solving ordinary differential equations of initial-value type. It could be improved a little by a Crank-Nicolson-type approach to reduce the local truncation error, and this would probably not affect the conclusion, proved by Meyer, that his approximation tends, as $\Delta t \to 0$, to a weak solution of the Stefan problem. We shall meet the "weak solution" again very soon.

Meyer also considers the inverse Stefan problem, where the moving boundary $s(t)$ is given, and we wish to determine some other function, say $h(t)$ or $g(t)$ in the second of (2.30). For the former we proceed as before to calculate W_k, and this gives

$$w_k(s_k) = \left\{\frac{s_k - s_{k-1}}{\Delta t\ m_k(s_k)} - \frac{\ell_k(s_k)}{m_k(s_k)}\right\} W_k(s_k) \ . \tag{2.36}$$

Then $h_k = w_k(0)$, where $w_k(s_k)$ is an initial condition for the backward integration of

$$w_k' = -\left(\frac{c_k}{\Delta t} + b_k\right) W_k w_k + \left(\frac{c_k}{\Delta t} u_{k-1} - f_k\right) W_k \ . \tag{2.37}$$

A similar but slightly more complicated procedure can determine $g(t)$ if $h(t)$ and $s(t)$ are given.

(b) *Change of independent variable*

The next important modification involves a shuffling in the rôles of the dependent and independent variables, and has apparently been proposed by several different workers. It is called by some the Isotherm Migration Method, and a recent account of it is given by Crank and Phahle [10].*

The general idea, of course, has been used for some time for the solution of problems in hydrodynamics. For example, in irrotational flow a curved free stream line on which ψ = constant in the (x,y) plane becomes a straight line in the (ϕ,ψ) plane, where ϕ and ψ are harmonic conjugates. Similarly,

---

* See also p. 201.

in our heat problem, we seek u as a function of x and t, and this is converted into a problem in which we determine x as a function of u and t.

Crank and Phahle [10] consider the problem given by

$$\left.\begin{aligned}
&\frac{\partial u}{\partial t} = \frac{\partial^2 u}{\partial x^2},\ 0 < x < s(t),\\
&u = u_0,\ x = 0,\\
&u = 0,\ 0 < x < 1,\ t = 0,\\
&u = 0,\ \text{and } S\,\frac{dx}{dt} = -\frac{\partial u}{\partial x} \text{ at } x = s(t)
\end{aligned}\right\}. \qquad (2.38)$$

The transformed equations are

$$\left.\begin{aligned}
&\frac{\partial x}{\partial t} = \left(\frac{\partial x}{\partial u}\right)^{-2} \frac{\partial^2 x}{\partial u^2}\\
&x = 0 \text{ on } u = u_0\\
\text{and}\quad &S\,\frac{dx}{dt} = -\left(\frac{\partial x}{\partial u}\right)^{-1},\ x = s(t) \text{ on } u = 0
\end{aligned}\right\}. \qquad (2.39)$$

The situation in the (u,t) plane is shown in Fig. 8, where $x = 0$ on the fixed line $u = u_0$, and we seek to determine $x = s(t)$ on the line $u = 0$. If a general mesh value is $x_i^n = x(i\Delta u,\ n\Delta t)$, we seek in fact the values of x_0^n.

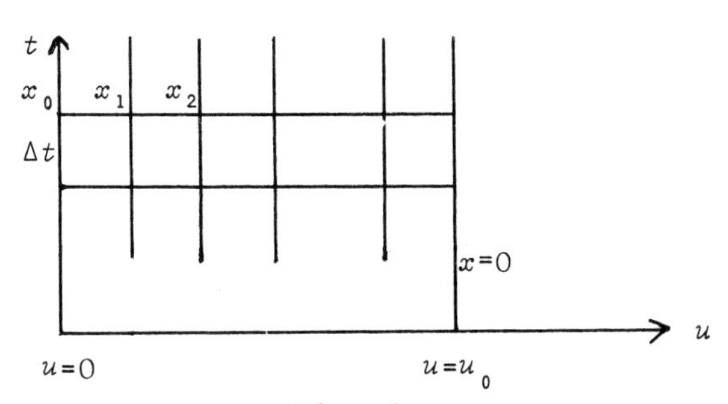

Fig. 8.

Crank and Phahle use the explicit finite-difference equations

$$\left.\begin{aligned}
&\frac{x_i^{n+1} - x_i^n}{\Delta t} = \frac{x_{i+1}^n - 2x_i^n + x_{i-1}^n}{\left(\dfrac{x_{i+1,n} - x_{i-1,n}}{2}\right)^2} \text{ for } i = 1,2,\ldots n-1\\
&\frac{S}{\Delta t}\left(x_0^{n+1} - x_0^n\right) = \frac{\Delta u}{x_1^n - x_0^n}
\end{aligned}\right\} \qquad (2.40)$$

and the computation is trivial.

The explicit method, as usual, might require a rather small Δt for stability, and the local truncation error in the second of (2.40) is rather larger than one would like in the region of most interest to us. The former disadvantage can be removed by using a Crank-Nicolson-type implicit scheme for both of (2.40), and although this gives rise to the solution of sets of nonlinear algebraic equations this is no major problem with modern computers and with the good start we should normally expect to have (if only by using the explicit scheme for this purpose) for the relevant iterative method. The second problem can be removed by introducing "fictitious points" and using central differences at the boundary $u = 0$, without introducing serious complication in the Crank-Nicolson nonlinear system.

This seems to me to be a major improvement in technique, and further developments, promised by Crank and Phahle, should be followed with interest. I doubt, however, if it could become the best method for the many varieties of problem which exist in this field, and that more than one space dimension, for example, could give real difficulties.

(*c*) *Change of differential equation*

Finally, in this discussion of finite-difference methods, I come to what seem to me to be the best general-purpose methods. Szekely and Themelis [21] consider the problem given by the equations (2.1) to (2.3) on $0 \leq x \leq \infty$. They remark that if H is the sum of the sensible and latent heat content of the material the differential equations can be replaced by the single equation

$$\frac{\partial}{\partial x}\left(k\,\frac{\partial u}{\partial x}\right) = \rho\,\frac{\partial H}{\partial t},\quad 0 \leq x \leq \infty, \tag{2.41}$$

where k, ρ and H are functions of u. At the interface $x = s(t)$ there are step changes in k and H, and they propose to replace these by continuous curves, that is take $k(u)$ and $H(u)$ as smooth functions of u, effectively producing a melting region rather than a melting point. They can then replace the system of equations by the single equation

$$\frac{\partial}{\partial x}\left(k\,\frac{\partial u}{\partial x}\right) = \rho\left(\frac{dH}{du}\right)\frac{\partial u}{\partial t}, \tag{2.42}$$

in which k, ρ and H are known functions of u.

In this approach the interface effectively disappears, and there is no need for special equations or special points in any time or space region. They use explicit finite-difference equations and easily account for the variable nature of the coefficients in (2.42).

Now this is obviously a step in the direction of the so-called "weak" solution, and I shall finally mention two recent papers in which the idea of a weak solution is exploited to give very convenient numerical methods.

Atthey [1]* considers the differential equation

$$k\,\frac{\partial^2 u}{\partial x^2} + Q = \rho\,\frac{\partial H}{\partial t}\,, \tag{2.43}$$

where Q is a known function of u and H has the definition of the previous paragraphs. Instead of an interface condition like

$$k_2\,\frac{\partial u_2}{\partial x} - k_1\,\frac{\partial u_1}{\partial x} = -\,\rho L\,\frac{ds}{dt}(t) \tag{2.44}$$

he takes

$$\left.\begin{aligned} u &= H/c, & H &\leqslant c\,u_m \\ u &= u_m, & c\,u_m &\leqslant H \leqslant c\,u_m + L \\ u &= (H-L)/c, & H &\geqslant c\,u_m + L \end{aligned}\right\} \tag{2.45}$$

where u_m is the melting temperature. He has virtually replaced the "smooth" H of Szekely and Themelis [21] by a piecewise continuous function, and as a result he proceeds to prove that the solution obtained by his particular numerical method converges, as the intervals tend to zero, to a weak solution of the differential equation.

---

* See also p. 182.

His method is based on the simplest explicit formula

$$\rho \frac{\left(H_i^{n+1} - H_i^n\right)}{\Delta t} = Q_i^n + \frac{k}{(\Delta x)^2}\left(u_{i+1}^n - 2u_i^n + u_{i-1}^n\right), \quad (2.46)$$

the known relations between H and u providing a simple numerical procedure. Atthey proves that as Δt, $\Delta x \to 0$ the computed results tend to the type of weak solution described in his paper, provided only that the numerical scheme is stable which implies, as is usual with explicit schemes, some restriction on the size of the time step-length Δt.

The same idea has been used by Meyer [19], though with implicit methods to avoid numerical instabilities. As I have indicated frequently throughout this paper this is certainly preferable in one space dimension, and even in several space dimensions with fast computers and good methods for solving elliptic equations.

Meyer considers the equation

$$\nabla^2 u = \frac{\partial H}{\partial t}, \quad (2.47)$$

and approximates to H by the piecewise continuous functions

$$\left.\begin{aligned} H(u) &= c_1 u, \quad u \leqslant u_m - \varepsilon \\ H(u) &= H(u_m - \varepsilon) + \frac{L}{2\varepsilon}(u - u_m + \varepsilon), \quad u_m - \varepsilon \leqslant u \leqslant u_m + \varepsilon \\ H(u) &= H(u_m + \varepsilon) + c_2(u - u_m - \varepsilon), \quad u \geqslant u_m + \varepsilon \end{aligned}\right\} \quad (2.48)$$

observing that the solid-liquid transition realistically occurs over a range $u_m - \varepsilon$, $u_m + \varepsilon$ of temperature.

With given initial and boundary conditions Meyer uses the implicit formula

$$\frac{H_i^{n+1} - H_i^n}{\Delta t} = \nabla^2 u_i^{n+1} \quad (2.49)$$

so that at each stage he has to solve an elliptic problem. This avoids the problem of instability, but I would have been prepared even to use the Crank-Nicolson formula

$$\frac{H_i^{n+1} - H_i^n}{\Delta t} = \tfrac{1}{2}\left(\nabla^2 u_i^{n+1} + \nabla^2 u_i^n\right), \quad (2.50)$$

which has a smaller truncation error in the t-direction and in which $\nabla^2 u_i^n$ is a known and easily computable function of space and time.

Like Atthey, Meyer proves that his numerical solutions converge to the unique weak solution, in which

$$\int_{\Omega_T} \{u\nabla^2\phi + H(u)\frac{\partial\phi}{\partial t}\}dx\,dt - \int_0^T \oint_{\partial D} f\frac{\partial\phi}{\partial n}ds\,dt + \int_D H\{f(x,0)\}\phi(x,0)\,dx = 0 \tag{2.51}$$

for arbitrary smooth test functions ϕ. Here D is the region with boundary ∂D, T is some fixed time limit, $\Omega_T = (0,T) \times D$, and

$$u = f(x,t) \text{ on } \partial D, \; u = f(x,0) \text{ for } t = 0. \tag{2.52}$$

Meyer also makes the reasonable assertion that his numerical method should be virtually independent of ε, and this seems to be verified by his numerical results on problems in two space dimensions. He used a finite-difference method, with which (2.49) reduces to the system of nonlinear algebraic equations given by

$$Au + \phi(u) = 0, \tag{2.53}$$

where u is the vector of required u^{n+1} values and $\phi(u)$ includes the quantities already known in (2.49). He solves this by Gauss-Seidel iteration with some convergence-accelerating parameters, the question of which of the three regions is inhabited by a particular u being determined as part of the process.

His method seemed to take little computing time (on a CDC 6400), but he is a little vague about the details of the computation and recommends some research in this area. One might think of block SOR, alternating-direction methods, and it could be that a suitably-adapted finite-element method might be the winner. In any case this seems to me to be by far the best of the finite-difference methods for any significant problem, and especially in more than one space dimension.

2.4 Importance of analytical solutions

I mentioned in my introduction that the best methods would usually be a combination of numerical and analytical methods. Numerical methods, as we have seen, are somewhat unsatisfactory in the neighbourhood of almost any sort of singularity in the functions we are trying to compute. In this respect the smoothing at the interface, or the use of weak solutions, will not usually solve the problems which commonly arise for small times. The most obvious such problem involves sudden heating or sudden cooling, and the point $x = 0$, $t = 0$ will be "difficult" for example when the boundary and initial conditions are

$$u(x,0) = 0, \quad u(0,t) = 1. \tag{2.54}$$

If we use "brute force" finite-difference methods, taking very small intervals to reduce and localise the effect of the singularity, then we may be performing a prohibitive amount of computation. Moreover, though with conditions like (2.54) the effects of the singularity will decrease with time, this may not be the case if the conditions are, for example, changed to

$$u(x,0) = 0, \quad \frac{\partial u}{\partial x}(0,t) = 1, \tag{2.55}$$

and there may be a "persistent" discretisation error as demonstrated by Crank and Parker [9].

In such circumstances a short-time analytical solution is the obvious answer, and such solutions are quite commonly obtainable even for complicated problems. They will, moreover, be sufficiently accurate for small times, and the solution can be conveniently extended to larger times by one or other of our numerical methods.

3. INTEGRAL AND INTEGRO-DIFFERENTIAL EQUATION METHODS

A number of workers have solved selected problems in terms of integro-differential equations. For example Boley [4]* considers a liquid region (1) in $0 < x < s(t)$, and a solid

---

* See also p. 150.

region (2) in $s(t) < x < \ell$, governed by the equations

$$\left.\begin{aligned} K\frac{\partial^2 u_1}{\partial x^2} &= \frac{\partial u_1}{\partial t} \\ k\frac{\partial u_1}{\partial x} + h\,u_1 &= f_1(t) \text{ at } x = 0 \end{aligned}\right\}, \qquad (3.1)$$

$$\left.\begin{aligned} K\frac{\partial^2 u_2}{\partial x^2} &= \frac{\partial u_2}{\partial t} \\ k_1\frac{\partial u_2}{\partial x} + h_1 u_2 &= 0 \text{ at } x = \ell \end{aligned}\right\}, \qquad (3.2)$$

$$\left.\begin{aligned} u_2(s(t),t) &= u_1(s(t),t) = u_m \\ k\frac{\partial u_2}{\partial x} + h\,u_m &= \rho L\frac{ds}{dt} + f_1(t) \end{aligned}\right\} \text{ at } x = s(t). \qquad (3.3)$$

Boley considers one variable u, which satisfies the conditions

$$\left.\begin{aligned} &K\frac{\partial^2 u}{\partial x^2} = \frac{\partial u}{\partial t}, \\ &k\frac{\partial u}{\partial x} + h\,u_m = \rho L\frac{ds}{dt} + f_1(t),\ u = u_m, \text{ at } x = s(t), \\ &k_1\frac{\partial u}{\partial x} + h_1 u = 0 \text{ at } x = \ell, \\ \text{and}\quad &k\frac{\partial u}{\partial x} + hu = f_1(t) + f(t) \text{ at } x = 0 \end{aligned}\right\}. \qquad (3.4)$$

The introduction of the unknown $f(t)$ makes all this possible, and we now have to determine $s(t)$ and $f(t)$. Using Duhamel's solution of the equations

$$\left.\begin{aligned} K\frac{\partial^2 u_0}{\partial x^2} &= \frac{\partial u_0}{\partial t},\ u_0(x,0) = 0 \\ k\frac{\partial u_0}{\partial x} + h\,u_0 &= 1 \text{ at } x = 0 \\ k_1\frac{\partial u_0}{\partial x} + h_1 u_0 &= 0 \text{ at } x = \ell \end{aligned}\right\}, \qquad (3.5)$$

the required functions are found to satisfy the simultaneous integro-differential equations

$$\left.\begin{aligned} &\int_0^t f_1(t-t_1)\frac{\partial u_0}{\partial t_1}(s(t),t_1)\,dt_1 + \int_0^{t-t_m} f(t-t_1)\frac{\partial u_0}{\partial t_1}(s(t),t_1)\,dt_1 = u_m \\ &k\int_0^t f_1(t-t_1)\frac{\partial^2 u_0}{\partial x\partial t_1}(s(t),t_1)\,dt_1 + k\int_0^{t-t_m} f(t-t_1)\frac{\partial^2 u_0}{\partial x\partial t_1}(s(t),t_1)\,dt_2 \\ &\qquad = \rho L\frac{ds}{dt}(t) + f_1(t) - hu_m \end{aligned}\right\} \qquad (3.6)$$

with $s(t_m) = 0$, where t_m is the time at which melting first starts. The temperature is then obtained easily from the equation

$$u(x,t) = \int_0^t f_1(t-t_1)\frac{\partial u_0}{\partial t_1}(x,t_1)dt_1 + \int_0^{t-t_m} f(t-t_1)\frac{\partial u_0}{\partial t_1}(x,t_1)dt_1. \quad (3.7)$$

It is here assumed that the melted portion is removed instantly, so that only the solid part is considered. Otherwise, for example in the solidification of a liquid contained between parallel walls, it may be necessary to include more unknown functions and to solve up to four simultaneous integro-differential equations, in which the thermal constants like K and k are different in the solid and liquid regions.

Boley solves problems in which solutions like u_0 can be written down immediately and analytically, and treats the integro-differential equations by analytical methods for short times and by numerical methods in the continuation to larger times.

Similar integro-differential equations were also produced by Kolodner [16] by a somewhat different method. With some simplifying conditions he considers the problem of the freezing of a lake of finite depth, with equations

$$\left.\begin{aligned} K_1 \frac{\partial^2 u_1}{\partial x^2} &= \frac{\partial u_1}{\partial t} \\ u_1(0,t) &= u_{1,0} \end{aligned}\right\}, \quad 0 < x < s(t) \qquad (3.8)$$

$$\left.\begin{aligned} K_2 \frac{\partial^2 u_2}{\partial x^2} &= \frac{\partial u_2}{\partial t} \\ u_2(x,0) &= u_{2,0} \\ \frac{\partial u_2}{\partial x}(h,t) &= 0 \end{aligned}\right\}, \quad \begin{aligned} s(t) &< x < h \\ 0 = s(0) &< x < h \end{aligned} \qquad (3.9)$$

and

$$\left.\begin{aligned} u_1 &= u_2 = 0 \\ k_1 \frac{\partial u_1}{\partial x} - k_2 \frac{\partial u_2}{\partial x} &= \rho L \frac{ds(t)}{dt} \end{aligned}\right\} \text{ at } x = s(t). \qquad (3.10)$$

He shows that the solution is obtained from the three simultaneous integro-differential equations

$$\left.\begin{aligned}
g_1(t) &= -\int_0^t \frac{g_1(\tau)}{t-\tau}\Big\{z_1(s,\tau)\eta(z_1(s,\tau)) \\
&\qquad +(z_1(-s,\tau)\eta(z_1,(-s,\tau)))\Big\}d\tau - \frac{2u_{1,0}}{\sqrt{K_1 t}}\eta(z_1(s,0) \\
g_2(t) &= \int_0^t \frac{g_2(\tau)}{t-\tau}\Big\{z_2(s,\tau)\eta(z_2(s,\tau))-z_2(2h-s,\tau)\eta(z_2(2h-s,\tau))\Big\}d\tau \\
&\qquad + \frac{u_{2,0}}{\sqrt{K_2 t}}\Big\{\eta(z_2(s,0))-\eta(z_2(2h-s,0))\Big\}, \\
\frac{ds}{dt} &= \frac{k_1}{\rho L}g_1(t) - \frac{k_2}{\rho L}g_2(t), \quad s(0) = 0
\end{aligned}\right\} \quad (3.11)$$

$$\text{where}\quad \left.\begin{aligned}
g_1(t) &= \frac{\partial u_1}{\partial x}(s(t),t), \quad g_2(t) = \frac{\partial u_2}{\partial x}(s(t),t), \\
z_1(x,\tau) &= \frac{x-s(t)}{2\sqrt{K_1(t-\tau)}}, \quad z_2(x,\tau) = \frac{x-s(t)}{2\sqrt{K_2(t-\tau)}}, \\
\eta(\xi) &= \frac{1}{\sqrt{\pi}}e^{-\xi^2}.
\end{aligned}\right\} \quad (3.12)$$

Integro-differential equations have also been applied by Hansen and Hougaard [14] to the implicit problem of Crank and Gupta, and despite the quoted forebodings of Sackett the approach appears to be successful*. For this problem they exploit a Green's function and produce for the moving boundary $s(t)$ an equation of the type

$$s = f_1(s,t) + \int_0^t f_2(s,s';t,t')dt' \qquad (3.13)$$

where $s = s(t)$, $s' = s(t')$, and f_1 and f_2 are somewhat complicated nonlinear expressions involving exponential and erfc functions. They solve this equation by series methods for small t and by numerical methods for larger t, following which the temperature is computed relatively easily from an expression involving functions and integrals of functions of $s(t)$.

The Green's function is also used by Chuang and Szekely [6], who consider the problem defined by the equations

$$\left.\begin{aligned}
&K\frac{\partial^2 u}{\partial x} - \frac{\partial u}{\partial t} = 0, \; -s(t) \leqslant x \leqslant s(t) \\
&u = u_0(x),\; t = 0;\; u = u_m(t),\; x = \pm s(t), \\
&\alpha\frac{\partial u}{\partial x} + \beta u = f(t),\; x = \pm s(t), \\
&s(t) = 1 \text{ at } t = 0
\end{aligned}\right\} \quad (3.14)$$

* See also Ockendon (p. 141).

where $f(t)$ is related to the movement of the phase boundary. With the use of the Green's function

$$G = \frac{1}{2\sqrt{K\pi(t-\tau)}} \exp\{-(x-\xi)^2/4K(t-\tau)\}, \tag{3.15}$$

they produce for the solution an integro-differential equation for u and $s(t)$, the form of $f(t)$ being known.

Now all these integro-differential equations are of Volterra type, and therefore probably amenable to numerical solution without too much difficulty, despite the obvious complications near $t = 0$. But as yet no relative evaluation has been made compared with the best finite-difference methods, and this is an obvious area for further research. All these methods, moreover, seem to depend on the ability to be able to write down analytical solutions for some parts of the problem, and they therefore probably lack the versatility of direct numerical approaches to the most general problems, in which, for example, the thermal properties are functions of temperature even within the liquid or solid regions.

Integral equations are also produced in the heat-balance integral method. Consider, for example, the problem defined by the equations

$$\left.\begin{aligned} K_1 \frac{\partial^2 u_1}{\partial x^2} &= \frac{\partial u_1}{\partial t}, 0 \leqslant x \leqslant s(t) \\ K_2 \frac{\partial^2 u_2}{\partial x^2} &= \frac{\partial u_2}{\partial t}, s(t) \leqslant x \leqslant \ell \end{aligned}\right\}, \tag{3.16}$$

with some condition for u_1 at $x = 0$ and for u_2 at $x = \ell$, and with interface conditions given by

$$k_1\frac{\partial u_1}{\partial x} - k_2\frac{\partial u_2}{\partial x} = \rho_2 L\frac{ds}{dt},\ u_1 = u_2 = u_m,\ \text{at } x = s(t). \tag{3.17}$$

The first step is to integrate the first of (3.16) from 0 to $s(t)$ and the second of (3.16) from $s(t)$ to ℓ, producing the integral equations

$$\left.\begin{aligned} K_1\left\{\left(\frac{\partial u_1}{\partial x}\right)_{s(t)} - \left(\frac{\partial u_1}{\partial x}\right)_0\right\} &= \int_0^{s(t)} \frac{\partial u_1}{\partial t}dx = \frac{d}{dt}\int_0^{s(t)} u_1 dx - u_m\frac{ds(t)}{dt} \\ K_2\left\{\left(\frac{\partial u_2}{\partial x}\right)_\ell - \left(\frac{\partial u_2}{\partial x}\right)_{s(t)}\right\} &= \int_{s(t)}^{\ell} \frac{\partial u_2}{\partial t}dx = \frac{d}{dt}\int_{s(t)}^{\ell} u_2 dx + u_m\frac{ds(t)}{dt} \end{aligned}\right\}. \tag{3.18}$$

The elimination of $\frac{\partial u_1}{\partial x}$ and $\frac{\partial u_2}{\partial x}$ at $x = s(t)$, from (3.18) and (3.17), then produces the equation

$$\left(-\rho_2 L - \frac{k_1}{K_1}u_m + \frac{k_2}{K_2}u_m\right)\frac{ds}{dt}(t) + \frac{k_1}{K_1}\frac{d\theta_1}{dt} + \frac{k_2}{K_2}\frac{d\theta_2}{dt} = k_2\left(\frac{\partial u_2}{\partial x}\right)_\ell - k_1\left(\frac{\partial u_1}{\partial x}\right)_0 \tag{3.19}$$

with $$\theta_1 = \int_0^{s(t)} u_1(x,t)dx, \quad \theta_2 = \int_{s(t)}^{\ell} u_2(x,t)dx. \tag{3.20}$$

Equation (3.19) has virtually eliminated any dependence on x, and this elimination is completed by assuming some forms for u_1 and u_2 such as

$$u_1(x,t) = a_1(t)+b_1(t)x+c_1(t)x^2, \quad u_2(x,t) = a_2(t)+b_2(t)x+c_2(t)x^2. \tag{3.21}$$

Satisfaction of all the boundary conditions and the interface conditions then make it possible to express the problem as the solution of a system of ordinary differential equations. For example, with boundary conditions

$$k_1\frac{\partial u_1}{\partial x} = -H \text{ at } x = 0, \quad \frac{\partial u_2}{\partial x} = 0 \text{ at } x = \ell, \text{ and } u_m = 0, \tag{3.22}$$

Goodman and Shea [13] produce the differential equations

$$\left.\begin{aligned} -\rho_2 L\frac{ds(t)}{dt} + \frac{k_1}{K_1}\frac{d\theta_1}{dt} + \frac{k_2}{K_2}\frac{d\theta_2}{dt} &= H \\ \theta_1 &= (s(t))^2\left(\frac{H}{2k_1} - \frac{L}{3K_1}\frac{d\theta_1}{dt}\right) \\ \theta_2 &= -\frac{(\ell-s)^2}{3K_2}\frac{d\theta_2}{dt} \end{aligned}\right\} \tag{3.23}$$

With suitable initial conditions (which may not be trivial to obtain) they proceed to solve these equations approximately by rather complicated analysis, but of course they are easily soluble by numerical methods of the Runge-Kutta type.

Of all the proposed methods this must be numerically the easiest, but some writers claim that the assumed temperature profiles must be selected with great care and that, for example, taking higher-order polynomial forms does not necessarily give better accuracy. There is clearly room for research on this point, and also to determine what problems are soluble by this heat-balance integral method. Goodman [12] claims that variable thermal properties are easily dealt with, but there must be a limitation on the complexity of problems which can be solved with satisfactory accuracy by this method, since otherwise the numerical method is here so simple that no other method need be contemplated!

4. MISCELLANEOUS METHODS

Finally, I mention briefly two methods which may not have useful general application, but which are obviously valuable and numerically easy for certain classes of problems.

4.1 Variational methods

Variational methods were applied by Biot and others in a variety of papers from about 1957 onwards. Typical is the paper by Biot and Daughaday [3]*, which solves the problem represented by Fig. 9.

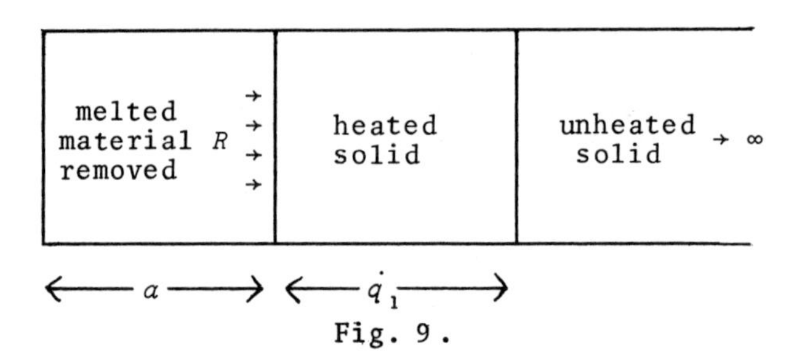

Fig. 9.

The Lagrangian equations are given by

$$\frac{\partial V}{\partial q_i} + \frac{\partial D}{\partial \dot{q}_i} = Q_i \tag{4.1}$$

where, in this example, the potential function V, dissipation function D and generalised thermal force Q_i are defined as

$$V = \tfrac{1}{2}\int_a^{a+q_1} cu^2\,dy, \quad D = \tfrac{1}{2}\int_a^{a+q_1} \frac{1}{k}\,\dot{H}_y^2\,dy, \quad Q_i = \left(u_a\,\frac{\partial H_y}{\partial q_i}\right)_{y=a}, \tag{4.2}$$

and k is the thermal conductivity, c the heat capacity, u the temperature rise, u_a the applied temperature on the boundary, and the heat flow distribution is

* See also Agrawal (p. 242).

$$H_y = \int_y^{a+q_1} cu \; dy. \tag{4.3}$$

The q_i are a convenient set of parameters describing the heat flow, and here just one is used, the penetration distance q_1.

It now again appears necessary to assume some sort of temperature profile, and Biot and Daughaday assume that

$$T = u_m\left\{1 - (y-a)/q_1\right\}^3, \tag{4.4}$$

where u_m (= u_a here) is the melting temperature. All these equations then produce one ordinary differential equation, given by

$$q_1\left(\frac{1}{28}\dot{q}_1 + \frac{11}{112}\dot{a}\right) = \frac{5}{14}\frac{k}{c}, \tag{4.5}$$

connecting the penetration distance with the rate of movement of the melting surface.

A second relation comes from the energy conservation requirement, given by

$$\begin{aligned} R &= (L + cu_m)\dot{a} + \frac{d}{dt}\int_a^{a+q_1} cudy \\ &= (L + cu_m)\dot{a} + \tfrac{1}{4}\, cu_m\, \dot{q}_i \end{aligned} \tag{4.6}$$

in this example. The solution of these two equations, with some assumed starting $q(0)$, follows easily.

In a later paper Biot and Agrawal [2] extended the treatment to the case when the physical properties of the material depend on the temperature. The analysis is somewhat more complicated, but the numerical work still involves only the solution of an ordinary differential equation.

4.2 Computation of upper and lower bounds

I have already mentioned the work of Boley [4], who produced and solved an integro-differential equation with the introduction of one or more "fictitious" functions. In Boley

[5] he uses the same idea, but instead of trying to solve the integral equation he finds upper and lower bounds for the solution. He considers the more general equations

$$\left.\begin{aligned} &\frac{\partial}{\partial x}\left(k\,\frac{\partial u}{\partial x}\right) = \rho c\,\frac{\partial u}{\partial t},\quad s(t) < x < \ell,\\ &u(x,0) = 0,\quad \frac{\partial u}{\partial x}(L,t) = 0,\\ &-k\,\frac{\partial u}{\partial x}(0,t) = Q(t)\ \text{(before melting starts at time } t_m \text{ with temperature } u_m), \end{aligned}\right\}\quad(4.7)$$

with interface conditions after melting starts given by

$$-k_m\frac{\partial u}{\partial x}\{s(t),t\} = Q(t) - \rho_m L\frac{ds}{dt},\quad u\{s(t),t\} = u_m,\quad t > t_m,\quad(4.8)$$

and where

$$\int_0^t Q(t)\,dt = \int_{s(t)}^{\ell} H(u)\,dx + (\rho_m L + H_m)s(t),\quad H(u) = \int_0^u \rho(u')c(u')\,du'.\quad(4.9)$$

The quantity $Q(t)$ is an arbitrarily prescribed heat input at $x = 0$, and after melting continues to be applied at $x = s(t)$.

Boley then shows that any solution of (4.7) also satisfies (4.8) for some particular $Q(t)$ and $s(t)$. He therefore writes down the fundamental solution $u_0(x,t)$ of (4.7), in the third of which $Q(t)$ is replaced by unity, and then observes that the solution for any heat input $Q^*(t)$ at $x = 0$, which is just $Q(t)$ for $t \leqslant t_m$, is given by

$$\left.\begin{aligned} u^*(x,t) &= \int_0^t Q(t_1)\frac{\partial u_0}{\partial t_1}(x,t-t_1)\,dt_1,\quad t \leqslant t_m\\ u^*(x,t) &= \int_0^{t_m} Q(t_1)\frac{\partial u_0}{\partial t_1}(x,t-t_1)\,dt_1 + \int_{t_m}^{t} Q^*(t_1)\frac{\partial u_0}{\partial t_1}(x,t-t_1)\,dt_1,\\ &\qquad t \geqslant t_m \end{aligned}\right\}.\quad(4.10)$$

He then computes u^* for selected choices of Q^*, and for each evaluates $s(t)$ from the second of (4.8) and $Q(t)$ from (4.9). Important theorems, proved in the paper, show that the melting rate $s(t)$ is lower than the true rate in $t_m \leqslant t \leqslant t^*$ if $Q^* \leqslant Q$ in that range, and higher if $Q^* > Q$ in that range.

The arithmetic is fairly straightforward, though it is not clear how to choose the Q^* so that the distance between the upper and lower bounds is satisfactorily small. Moreover, as usual with semi-analytical methods, the process fails unless the standard $u_0(x,t)$ can be written down immediately and Duhamel's theorem applies.

4.3 Perturbation methods

Finally, I mention a class of perturbation methods in which attempts are made to solve the given differential system by means of series expansions. Typical of these is the paper by Jiji [15], who considers a solid cylinder at temperature u_0 surrounded with fluid originally at temperature u_i for which the solidification temperature is u_f. For $u_0 < u_f < u_i$ solidification starts at the surface of the cylinder. With the introduction of certain non-dimensional parameters the equations for solution become

$$\left.\begin{aligned}
\frac{\partial u}{\partial t} &= \frac{\partial^2 u}{\partial \xi^2} + \frac{1}{\xi}\frac{\partial u}{\partial \xi}, \quad 0 < \xi < 1 \quad \text{(cylinder)}\\
\frac{\partial v}{\partial t} &= \frac{\alpha_2}{\alpha_1}\left(\frac{\partial^2 v}{\partial \xi^2} + \frac{1}{\xi}\frac{\partial v}{\partial \xi}\right), \quad 1 < \xi < 1 + \sigma \quad \text{(solidified fluid)}\\
\frac{\partial w}{\partial t} &= \frac{\alpha_3}{\alpha_1}\left(\frac{\partial^2 w}{\partial \xi^2} + \frac{1}{\xi}\frac{\partial w}{\partial \xi}\right), \quad 1 + \sigma < \xi < \infty \quad \text{(fluid)}
\end{aligned}\right\} \quad (4.11)$$

with initial, boundary and interface conditions given by

$$\left.\begin{aligned}
&u(\xi,0) = -1,\ w(\xi,0) = w_i,\ u(1,\tau) = \frac{k_1}{k_2}v(1,\tau), \frac{\partial u}{\partial \xi}(1,\tau) = \frac{\partial v}{\partial \xi}(1,\tau)\\
&v(1+\sigma,\tau) = w(1+\sigma,\tau) = 0,\ w(\infty,\tau) = w_i\\
&\frac{d\sigma}{d\tau} = \varepsilon\left\{\frac{\partial v}{\partial \xi}(1+\sigma,\tau) - \frac{\partial w}{\partial \xi}(1+\sigma,\tau)\right\}
\end{aligned}\right\} \quad (4.12)$$

where ε is a multiple of $u_f - u_0$.

Jiji seeks approximate solutions in terms of series expansions such as

$$\left.\begin{aligned} u(\xi, \tau; \varepsilon) &= \sum_{n=0}^{\infty} \varepsilon^n u_n(\xi, \tau) \text{ (and similarly for } v \text{ and } w) \\ \sigma(\tau, \varepsilon) &= \sum_{n=0}^{\infty} \varepsilon^n \sigma_n(\tau) \end{aligned}\right\} \quad (4.13)$$

Not surprisingly, he finds that such an expansion is not possible for v, and has to change the variables to

$$x = \frac{\xi - 1}{\varepsilon}, \quad \bar{\sigma}(\tau, \varepsilon) = \frac{\sigma(\tau, \varepsilon)}{\varepsilon}, \quad \bar{v}(x, \tau; \varepsilon) = \frac{v(\xi, \tau; \varepsilon)}{\varepsilon}. \quad (4.14)$$

He now uses series of type (4.13) for $\bar{v}$ and $\bar{\sigma}$, and after expressing all the relevant equations as power series in ε he proceeds to find equations for successive approximations obtained by equating to zero the coefficients of ε^0, ε^1, $\varepsilon^2, \ldots$.

The analysis is very complicated, and even so Jiji finds it possible to obtain only the zero-order approximation. It would clearly be very difficult to use this method with guaranteed success on more complicated problems and especially in more than one space dimension.

5. CONCLUSION

As I said in the introduction, there is no single "best" method, and I can reach only a few main conclusions. First, I believe that the best all-purpose method, to cope with complicated problems with variable thermal properties and with several dimensions, must be the numerical method applied to the original equations modified to produce a weak solution. This I should take to be of implicit type, with the solution of a boundary-value problem at each time step. Some research is needed here to decide on the relative performance of finite-difference and finite-element methods, and on appropriate methods for solving the nonlinear equations which either method would produce.

Second, I think that the heat-balance integral method is the easiest to apply, and for the limited range of problems for which it is applicable and sufficiently accurate both the analysis and the arithmetic are so relatively easy that this must be a very competitive method. A strong competitor for problems in one space dimension would be the Meyer method of solving two-point boundary-value problems, at each time step, by turning them into initial-value problems, and this method also appears competent to cope with the inverse Stefan problem. Another possible competitor, but one for which more research is needed, is the isotherm-migration method.

I have not been able to get any real idea about the relative merits of the use of integro-differential equations, but it is clear that doubts must be expressed about the claims to be general-purpose methods of all techniques which make some simplifying assumption, for example about the nature of the temperature profile, and even on techniques which depend for their success on the ability to write down exact solutions in some part of the analysis.

I would therefore suggest that analytical research should concentrate on an attempt to determine conditions under which all these various assumptions and requirements are more or less valid. Much of the literature I have read and quoted here has spent much time in discussing what are virtually unnecessary topics, such as how to solve a system of nonlinear differential equations. These are so easily solved by numerical methods that such analysis is not only unnecessary but even counter-productive, giving rise to incorrect evaluations of the various methods.

I conclude that we already have some useful and valuable knowledge, but that considerable research is still needed of both analytical and numerical kinds.

REFERENCES

1. Atthey, D.R., "A finite-difference scheme for melting problems," *J. Inst. Math. App.*, **13**, 353-366 (1974).
2. Biot, M.A. and Agrawal, H.C., "Variational analysis of ablation for variable properties," *J. Heat Transfer* **86**, 437-442 (1964).

3. Biot, M.A. and Daughaday, H., "Variational analysis of ablation," *J. Aerospace Sci.* 29, 227-229 (1962).

4. Boley, B.A., "A method of heat-conduction analysis of melting and solidification problems," *J. Math. Phys.* 40, 300-313 (1961).

5. Boley, B.A., "Upper and lower bounds for the solution of a melting problem," *Q. App. Math.* 21, 1-11 (1963).

6. Chuang, Y.K. and Szekely, J., "On the use of Green's functions for solving melting or solidification problems," *Int. J. Heat Mass Transfer,* 14, 1285-1294 (1971).

7. Crank, J. and Gupta, R.S., "A moving boundary problem arising from the diffusion of oxygen in absorbing tissue," *J. Inst. Maths. Applics.* 10, 19-33 (1972).

8. Crank, J. and Gupta, R.S., "A method for solving moving boundary problems in heat flow using cubic splines or polynomials," *J. Inst. Maths. Applics.* 10, 296-304 (1972).

9. Crank, J. and Parker, I.B., "Persistent discretization errors in partial differential equations of parabolic type", *Comp. J.*, 7, 163-167 (1964).

10. Crank, J. and Phahle, R.D., "Melting ice by the isotherm migration method," *Bull. Inst. Maths. Applics.* 9, 12-14 (1973).

11. Douglas, J. and Gallie, T.M., "On the numerical integration of a parabolic differential equation subject to a moving boundary condition," *Duke Math. J.* 22, 557-571 (1955).

12. Goodman, T.R., "The heat balance integral - further considerations and refinements," *J. Heat Transfer,* 83, 83-86 (1961).

13. Goodman, T.R. and Shea, J.J., "The melting of finite slabs," *J. App. Mech.* 27, 16-24 (1960).

14. Hansen, E. and Hougaard, P., "On a moving boundary problem from biomechanics," *J. Inst. Math. Apps.,* 13, 385 -398 (197

15. Jiji, L.M., "On the application of perturbation to free-boundary problems in radial systems," *J. Franklin Inst.* 289, 281-291 (1970).

16. Kolodner, I.J., "Free boundary problem for the heat equation with applications to problems of change of phase," *Comm. Pure Appl. Math.* **9**, 1-31 (1956).

17. Lazaridis, A., "A numerical solution of the multidimensional solidification (or melting) problem," *Int. J. Heat Mass Transfer,* **13**, 1459-1477 (1970).

18. Meyer, G.H., "On a free interface problem for linear ordinary differential equations and the one-phase Stefan problem," *Num. Math.* **16**, 248-267 (1970).

19. Meyer, G.H., "Multidimensional Stefan problems," *SIAM J. Numer. Anal.* **10**, 522-538 (1973).

20. Sackett, G.G., "An implicit free boundary problem for the heat equation," *SIAM J. Numer. Anal.* **8**, 80-96 (1971).

21. Szekely, J. and Themelis, M.J., "Heat chapter with change of phase," Ch. 10 of "Rate phenomena in process Metallurgy," Wiley-Interscience (1971).

BIOT'S VARIATIONAL PRINCIPLE FOR MOVING BOUNDARY PROBLEMS

H. C. Agrawal
(Pilkington Brothers Limited)*

1. INTRODUCTION

The fact that the only explicit closed form solution of a Stefan problem is that of Neumann (see Tayler, p. 127) highlights the importance of approximate analytical methods which can be applied when the boundary conditions are more complicated than just the prescription of the temperature, or when the material has temperature-dependent thermal properties. The heat balance integral method and variational methods fall into this category and the purpose of the present note is to describe briefly Biot's variational principle and its application to a melting problem with a radiation boundary condition.

2. BIOT'S VARIATIONAL PRINCIPLE

Variational principles based on the traditional concepts of the calculus of variations are well known, where a functional is extremised to determine an unknown function. Biot's variational principle on the other hand is derived on the basis of irreversible thermodynamic arguments applied to the physical problem. This formulation leads to a system of ordinary differential equations of Lagrangian type for a set of suitably chosen generalised coordinates for the system. The principle has been applied to a very wide class of phenomena (see Biot [2] for a full account) but the essential features for a thermal problem are

(*i*) the introduction of a heat flow vector $\underset{\sim}{H}\ (x,y,z,t)$,

---

* Permanent address: Department of Mechanical Engineering, Indian Institute of Technology, Kanpur, India.

(*ii*) the concepts of thermal potential, dissipation function and generalised thermal force.

We now describe the formulation of the variational principle for the energy equation following Biot [1]. Energy conservation for an isotropic solid with temperature u, thermal conductivity k and specific heat c gives

$$cu = -\mathrm{div}\ \underset{\sim}{H} \tag{2.1}$$

where the heat flow vector $\underset{\sim}{H}$ is such that the conduction law

$$\frac{\partial \underset{\sim}{H}}{\partial t} = \dot{\underset{\sim}{H}} = -k\ \underset{\sim}{\nabla}\ u \tag{2.2}$$

is satisfied. The "thermal potential" V is defined as

$$V = \tfrac{1}{2}\int_{\tau} cu^2\ d\tau \tag{2.3}$$

and for a given variation $\delta\underset{\sim}{H}$ we define

$$\delta D = \int_{\tau} \frac{1}{k}\ \dot{\underset{\sim}{H}} \cdot \delta\underset{\sim}{H}\ d\tau \tag{2.4}$$

The volume integrals in (2.3) and (2.4) are taken throughout a particular phase of the material at a particular time. Biot's variational principle may then be stated as

$$\delta V + \delta D = \int_{S} u\ \underset{\sim}{n} \cdot \delta\underset{\sim}{H}\ dS \tag{2.5}$$

where the surface integral is taken over the boundary of the material and $\underset{\sim}{n}$ is the unit inward normal. It can be interpreted as the generalised virtual work, $u\underset{\sim}{n}$ representing a force and $\delta\underset{\sim}{H}$ a virtual heat flow displacement. The physical motivation for (2.5) will be discussed further in Section 4.

The equivalence of (2.1) and (2.5) to the heat conduction equation can be seen by writing

$$\begin{aligned}\delta V &= \int_{\tau} cu\ \delta u\ d\tau = -\int_{\tau} u\ \mathrm{div}\ (\delta\underset{\sim}{H})\ d\tau \\ &= \int_{\tau} \delta\underset{\sim}{H} \cdot \underset{\sim}{\nabla} u\ d\tau + \int_{S} u\underset{\sim}{n} \cdot \delta\underset{\sim}{H}\ dS\end{aligned} \tag{2.6}$$

so that (2.5) gives

$$\int_\tau (\underset{\sim}{\nabla} u + \frac{1}{k}\dot{\underset{\sim}{H}}) . \delta\underset{\sim}{H}\, d\tau = 0. \tag{2.7}$$

Thus, if (2.5) holds for all $\delta\underset{\sim}{H}$, the heat conduction law (2.2) is obtained. The heat conduction equation

$$c\frac{\partial u}{\partial t} = \text{div}\ (k\ \underset{\sim}{\nabla}\ u) \tag{2.8}$$

may then be retrieved using (2.1).

This formulation can be generalised to include heat sources and temperature dependent thermal properties and, in particular to heat flow in a moving fluid with dissipation [4].

3. LAGRANGIAN FORMULATION

We write the heat flow vector $\underset{\sim}{H}$ as $\underset{\sim}{H}\ (q_i\ ;\ x,y,z)$ where $q_i(t)$ are generalised coordinates defining the thermal field. For example, we could write the temperature distribution in a bar maintained at temperature u_0 at $x = 0$ and insulated at $x = \ell$ as

$$c(u-u_0) = \sum_{n=0}^{\infty} q_n \sin\ (n+\tfrac{1}{2})\ \pi\frac{x}{\ell}\ , \tag{3.1}$$

when

$$H = \sum_{n=0}^{\infty} \frac{\ell q_n}{(n+\frac{1}{2})\pi} \cos\ (n+\tfrac{1}{2})\ \pi\frac{x}{\ell}\ . \tag{3.2}$$

In general,

$$\dot{\underset{\sim}{H}} = \sum_i \frac{\partial \underset{\sim}{H}}{\partial q_i} \dot{q}_i \tag{3.3}$$

and

$$\frac{\partial \dot{\underset{\sim}{H}}}{\partial \dot{q}_i} = \frac{\partial \underset{\sim}{H}}{\partial q_i} \tag{3.4}$$

so that

$$\delta\underset{\sim}{H} = \sum_i \frac{\partial \dot{\underset{\sim}{H}}}{\partial \dot{q}_i}\ \delta q_i\ . \tag{3.5}$$

Thus, from (2.4),

$$\delta D = \sum_i \delta q_i \int_\tau \frac{1}{k} \dot{\tilde{H}} \cdot \frac{\partial \dot{\tilde{H}}}{\partial \dot{q}_i} \, d\tau$$

$$= \sum_i \frac{\partial D}{\partial \dot{q}_i} \delta q_i \tag{3.6}$$

where

$$D = \int_\tau \frac{1}{2k} \dot{\tilde{H}}^2 \, d\tau \tag{3.7}$$

is called the "dissipation function".
In virtue of (3.5)

$$\int_S u\tilde{n} . \delta \tilde{H} dS = \sum_i Q_i \, \delta q_i \tag{3.8}$$

where

$$Q_i = \int_S u\tilde{n} \cdot \frac{\partial \tilde{H}}{\partial q_i} dS = \int_S u \frac{\partial H_n}{\partial q_i} dS \tag{3.9}$$

is the "thermal force" at the boundary (which need not be stationary). The "work" is done on a thermal system by a temperature u on a virtual heat displacement $\delta\tilde{H}$. Finally, since

$$\delta V = \sum_i \frac{\partial V}{\partial q_i} \delta q_i \, ,$$

the variational principle (2.5), with (3.6) and (3.8), gives

$$\sum_i \left\{ \frac{\partial V}{\partial q_i} + \frac{\partial D}{\partial \dot{q}_i} - Q_i \right\} \delta q_i = 0 \, . \tag{3.10}$$

If this equation holds for arbitrary δq_i we have the Lagrange equations

$$\frac{\partial V}{\partial q_i} + \frac{\partial D}{\partial \dot{q}_i} = Q_i \tag{3.11}$$

and the motivation for the nomenclature for V, D and Q is now apparent.

The solution of the problem now proceeds by assuming u to be a suitable function of q_i, x, y, z which satisfies the boundary conditions of the problem as in (3.1). The functions

V, D and Q_i are calculated from (2.3), (3.7) and (3.9), and (3.11) is then solved for $q_i(t)$. For example, for (3.1)

$$V = \frac{\ell}{4c} \sum_n q_n^{\,2} \;, \qquad D = \frac{\ell^3}{4k\pi^2} \sum_n \frac{\dot{q}_n^{\,2}}{(n+\frac{1}{2})^2} \tag{3.12}$$

and

$$Q_n \, \delta q_n = \frac{u_0 l}{(n+\frac{1}{2})\pi} \, \delta q_n \, . \tag{3.13}$$

Thus (3.11) gives

$$\frac{\ell^2}{2k\pi^2 (n+\frac{1}{2})^2} \, \dot{q}_n + \frac{1}{2c} \, q_n = \frac{u_0}{\pi (n+\frac{1}{2})} \tag{3.14}$$

in accordance with the exact solution of the problem. In general, of course, the use of an infinite number of generalised coordinates will lead to an intractable set of Lagrangian equations and our approach will therefore be to seek approximate solutions by taking u to be a function of only a finite number of generalised coordinates.

4. THE VARIATIONAL PRINCIPLE AS A MINIMUM DISSIPATION PRINCIPLE

The physical significance of the variational principle (2.5) can be seen by writing

$$P = D - \sum_i X_i \, \dot{q}_i \tag{4.1}$$

where

$$X_i = Q_i - \frac{\partial V}{\partial q_i} \, . \tag{4.2}$$

Minimising P as a function of $\dot{q}_i$, then gives

$$\frac{\partial D}{\partial \dot{q}_i} - X_i = 0 \tag{4.3}$$

which is (3.11). Now, from (3.9) and (2.6)

$$\sum_i X_i \, \dot{q}_i = \sum_i \dot{q}_i \left[\int_s u \underset{\sim}{n} \cdot \frac{\partial \underset{\sim}{H}}{\partial q_i} \, dS - \int_\tau \frac{\partial \underset{\sim}{H}}{\partial q_i} \cdot \underset{\sim}{\nabla} u \, d\tau - \int_s u \underset{\sim}{n} \cdot \frac{\partial \underset{\sim}{H}}{\partial q_i} \, dS \right]$$

$$= - \int_\tau \dot{\underset{\sim}{H}} \cdot \underset{\sim}{\nabla} u \, d\tau \tag{4.4}$$

so that (4.1) becomes

$$P = \frac{1}{2} \int_\tau \frac{1}{k} \dot{\underset{\sim}{H}}^2 \, d\tau + \int_\tau \dot{\underset{\sim}{H}} \, . \, \underset{\sim}{\nabla} u \, d\tau \, . \tag{4.5}$$

Now minimising P as a function of $\dot{q}_i$ is equivalent to minimising P as a function of $\dot{\underset{\sim}{H}}$ which, from (4.5), immediately gives the conduction law (2.2). We note that this also amounts to minimising D under the constraint

$$\int \dot{\underset{\sim}{H}} \, . \, \underset{\sim}{\nabla} u \, d\tau = \text{const.} \tag{4.6}$$

so that $\underset{\sim}{\nabla} u$ plays the rôle of a local disequilibrium force in a thermal system.

Biot's variational principle is thus a minimum dissipation principle and is an outgrowth of strong physical reasoning rather than just a mathematical formalism.

5. APPLICATION TO MELTING PROBLEMS

The variational principle has been used successfully to solve heat transfer problems with moving boundaries in systems having constant or temperature-dependent thermal properties [2], [5] and [6]. The volume integrals in the definitions of V and D have moving boundaries and the thermal force is defined on a moving surface.

We consider the one-dimensional example of aerodynamic heating at $x = 0$ of a solid in $x > 0$ initially at its melting temperature $u = 0$. Assuming constant thermal properties.

$$u_{xx} = \frac{1}{\beta} u_t \, , \quad 0 < x < s(t) \tag{5.1}$$

and

$$-ku_x = h_c(u_f - q(t)) \, , \; u_f > q \tag{5.2}$$

at $x = 0$ where u_f is the ambient temperature, $q(t) = u(0,t)$ and h_c is the heat transfer coefficient. At the interface $x = s$,

$$-ku_x = \rho L \dot{s}, \qquad u = 0 \tag{5.3}$$

and so

$$\dot{H} = \rho L\dot{s}. \tag{5.4}$$

The simplest possible approximation for u is a linear profile

$$u = q(1 - \frac{x}{s}) \tag{5.5}$$

which gives the correct temperature at $x = 0$ and $x = s$. We could relate q and s by (5.2) but for simplicity we initially regard q as given and s as the only generalised coordinate in writing down the Lagrangian equations. Only subsequently will we use (5.2) to relate s to q. The energy conservation (2.1), in one dimension, gives

$$\frac{\partial H}{\partial x} = - cu,$$

$$H = \rho Ls + \tfrac{1}{2}cqs(1 - \frac{x}{s})^2 \tag{5.6}$$

which satisfies (5.4). Then

$$V = \tfrac{1}{2}\int_0^s cu^2\, dx = \frac{1}{6}cq^2 s\ , \tag{5.7}$$

$$D = \frac{1}{2k}\int_0^s \dot{H}^2 dx = \frac{1}{2k}\left[\frac{c^2 s}{20}(q\dot{s} + \dot{q}s)^2 + \frac{c^2q^2s\dot{s}^2}{30} + \rho^2L^2s\dot{s}^2 + (q\dot{s} + \dot{q}s)\left\{\frac{c^2qs\dot{s}}{20} + \frac{c\rho Ls\dot{s}}{3}\right\} + \frac{c\rho Lqs\dot{s}^2}{3}\right], \tag{5.8}$$

$$Q = \left[u\frac{\partial H}{\partial s}\right]_{x=0} = \frac{cq^2}{2} + \rho Lq \tag{5.9}$$

Substituting for q from the relation $s = \dfrac{kq}{h_c(u_f-q)}$, obtained from (5.2), the Lagrangian equation $\dfrac{\partial V}{\partial s} + \dfrac{\partial D}{\partial \dot{s}} = Q$, in the generalised coordinate s, becomes

$$\dot{\eta}\,[A_1\eta^3 + A_2\eta^2 + A_3\eta + A_4u] = B_1\eta^2 + B_2\eta + B_3 \tag{5.10}$$

where $\eta = \dfrac{h_c s}{k}$ represents the dimensionless melting distance and A_i, B_i are suitable constants. Equation (5.10) can be solved explicitly and the melting distance η and the dimensionless surface temperature $\psi(=\frac{q}{u_f}) = \frac{\eta}{1+\eta}$ are plotted against dimensionless time $\zeta = \frac{h_c^2 u_f t}{\rho L k}$ in Figs. 1 and 2, respectively.

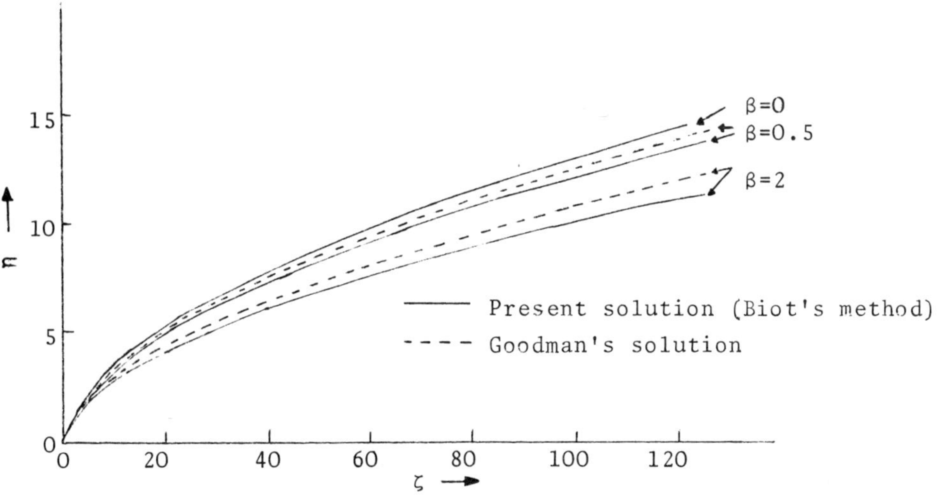

Fig. 1. Nondimensional melting distance.

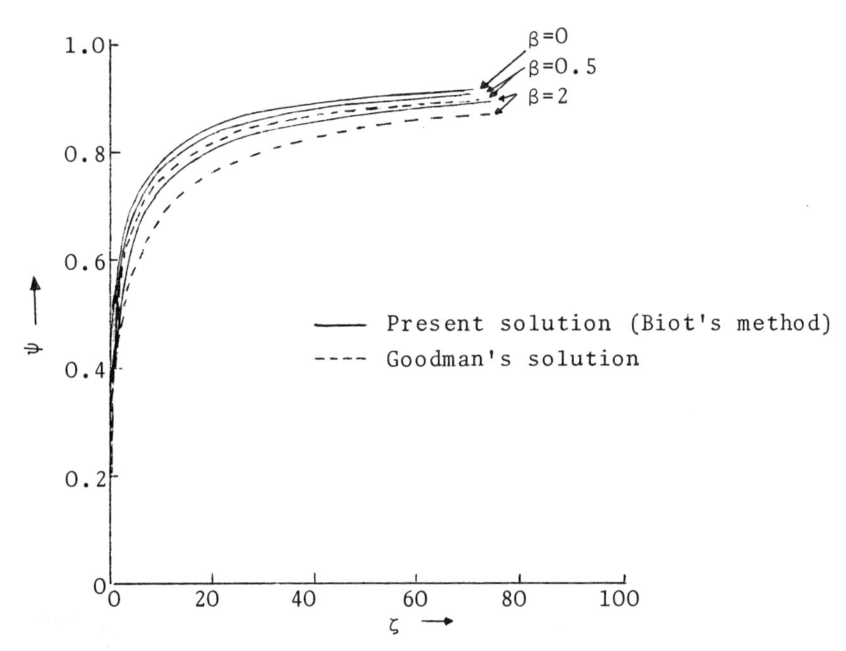

Fig. 2. Nondimensional surface temperature.

It is seen that good agreement is achieved with Goodman's [3] heat balance solution. Indeed, as $L \to \infty$, it is found that $\zeta \to \frac{\eta^2}{2} + \eta$ which is Goodman's result.

Further confirmation of the accuracy of the approximate variational solution has been made in the case when a constant heat flux is applied at $x = 0$, [5]. It is concluded that the variational approach offers a simple yet accurate method for obtaining approximate analytical solutions for phase change problems.

ACKNOWLEDGEMENTS

This paper is published with permission of the Directors of Pilkington Brothers Limited, and Dr. D. S. Oliver, Director of Group Research and Development.

REFERENCES

1. Biot, M.A., *J. Aeronautical Sci.*, **24**, 857-873 (1957).
2. Biot, M.A., "Variational Principles in Heat Transfer", Oxford U.P. (1970).
3. Goodman, T.R., *Trans. ASME*, **80**, 3, 335-342 (1958).
4. Nigam, S.D. and Agrawal, H.C., *J. Math. Mech.*, **9**, 869-884 (1960).
5. Prasad A. and Agrawal, H.C., *AIAA J*, **10**, 3, 325-327 (1972).
6. Prasad, A. and Agrawal, H.C., *AIAA J*, **12**, 250-252 (1974).

Fixation of a Moving Boundary by Means of a Change of Independent Variable

D. H. Ferriss

(National Physical Laboratory)

It is frequently the case that finite-difference methods are more straightforward in application to problems in nonlinear partial differential equations in regions of fixed extent than they are to linear equations in changing domains. Consequently, a particularly simple and useful preliminary device for the numerical solution of one-dimensional moving boundary problems in diffusion is the use of a change of space variable in order to fix the position of the moving boundary.* Under the transformation, the usual linear diffusion equation becomes a nonlinear partial differential equation containing terms involving the position of the moving boundary and its time derivative. The numerical solution of this nonlinear equation generally presents no serious difficulties and has the advantage that no special numerical techniques are necessary in the vicinity of the unknown boundary as would have been the case when working in the original space variable.

This technique was suggested, although not utilised, by Landau [3] for a one-phase ablation problem and has been employed by Crank [1] who considered a one-phase problem and also the application of two transformations to a two-phase problem.

We will consider the application of this technique to the Crank-Gupta oxygen diffusion problem discussed elsewhere in these proceedings, the equation for which, together with the initial and boundary conditions, are set out on p. 198.

---

*See Fox (p.216).

It is convenient to use first a change of dependent variable to make the partial differential equation homogeneous; we use

$$\phi = c + t \tag{1}$$

so that the equation becomes

$$\frac{\partial \phi}{\partial t} = \frac{\partial^2 \phi}{\partial x^2}, 0 < x < x_0(t) \tag{2}$$

with

$$\frac{\partial \phi}{\partial x} = 0 \text{ at } x = 0, \ t > 0, \tag{3}$$

$$\frac{\partial \phi}{\partial x} = 0 \text{ and } \phi = t \text{ at } x = x_0(t), \ t \geqslant 0 \tag{4}$$

and

$$\phi = \tfrac{1}{2}(1 - x)^2, \ 0 \leqslant x \leqslant 1, \ t = 0 \, . \tag{5}$$

Transformation to the independent variable

$$\eta = \frac{x}{x_0(t)}$$

leads to the relations

$$\left(\frac{\partial \phi}{\partial x}\right)_t = \frac{1}{x_0}\left(\frac{\partial \phi}{\partial \eta}\right)_t, \left(\frac{\partial^2 \phi}{\partial x^2}\right)_t = \frac{1}{x_0{}^2}\left(\frac{\partial^2 \phi}{\partial \eta^2}\right)_t,$$

$$\left(\frac{\partial \phi}{\partial t}\right)_x = \left(\frac{\partial \phi}{\partial t}\right)_\eta + \left(\frac{\partial \phi}{\partial \eta}\right)_t \left(\frac{\partial \eta}{\partial t}\right)_x = \left(\frac{\partial \phi}{\partial t}\right)_\eta - \frac{\eta}{x_0} \cdot \frac{dx_0}{dt}\left(\frac{\partial \phi}{\partial \eta}\right)_t .$$

Equation (2) becomes

$$\frac{\partial^2 \phi}{\partial \eta^2} = X \frac{\partial \phi}{\partial t} - \frac{\eta}{2}\frac{\partial \phi}{\partial \eta} \cdot \frac{dX}{dt} \tag{6}$$

where $X(t) = x_0{}^2(t)$, and conditions (3) - (5) become

$$\frac{\partial \phi}{\partial \eta} = 0 \quad \text{at } \eta = 0, t > 0, \tag{7}$$

$$\phi = t \text{ and } \frac{\partial \phi}{\partial \eta} = 0 \text{ at } \eta = 1, \ t \geqslant 0, \tag{8}$$

and $$\phi = \tfrac{1}{2}(1 - \eta)^2 \text{ for } 0 \leqslant \eta \leqslant 1, \text{ when } t = 0. \tag{9}$$

The derivatives in (6) are approximated by the finite-difference expressions

$$\left(\frac{\partial\phi}{\partial\eta}\right)_{i,j+\frac{1}{2}} = \frac{1}{4\delta\eta}\left\{(\phi_{i+1,j+1}-\phi_{i-1,j+1}) + (\phi_{i+1,j}-\phi_{i-1,j})\right\},$$

$$\left(\frac{\partial^2\phi}{\partial\eta^2}\right)_{i,j+\frac{1}{2}} = \frac{1}{2\delta\eta^2}\left\{(\phi_{i+1,j+1}-2\phi_{i,j+1}+\phi_{i-1,j+1}) + (\phi_{i+1,j}-2\phi_{i,j}+\phi_{i-1,j})\right\}$$

and $$\left(\frac{\partial\phi}{\partial t}\right)_{i,j+\frac{1}{2}} = \frac{\phi_{i,j+1}-\phi_{i,j}}{\delta t}\ .$$

Here $\eta = i\delta\eta$ and $t = j\delta t$ with $i = 0, 1, 2..N$, such that $N\delta\eta = 1$, and $j = 0, 1, 2......$ The quantity X in (6) is approximated by $\frac{1}{2}(X+X_0)$ where X_0 is the known value of X at the previous time level, and X now denotes the unknown value at the new time level. Finally, the rate of change of X is approximated using

$$\frac{dX}{dt} = \frac{X-X_0}{\delta t}\ .$$

Substitution of these approximations in (6) leads to the following relationship between the profiles at the jth and $(j+1)$th time levels.

$$\frac{A}{2}\phi_{i+1,j+1} + B\phi_{i,j+1} + \frac{C}{2}\phi_{i-1,j+1}$$

$$= -\frac{A}{2}\phi_{i+1,j} + D\phi_{i,j} - \frac{C}{2}\phi_{i-1,j} = R_i \tag{10}$$

where

$$A = \delta t + \frac{\eta}{4}(X-X_0)\delta\eta\ ,$$

$$B = -\ \delta t - \frac{\delta\eta^2}{2}(X+X_0)\ ,$$

$$C = \delta t - \frac{\eta}{4}(X-X_0)\delta\eta\ ,$$

$$D = \delta t - \frac{\delta\eta^2}{2}(X+X_0)\ .$$

The initial profile at $j=0$ is given by the condition (9). Use of the finite-difference form of the boundary condition (7) at $\eta=0$ and, for example, the second of (8) at $\eta=1$ together with (10), produces a system of equations

$$\underline{M}(\underline{X})\underline{\phi} = \underline{R} \tag{11}$$

where $\underline{M}(X)$ is a tridiagonal matrix whose elements depend on X:-

$$\begin{bmatrix} B & \frac{1}{2}(A_0+C_0) & & & & \\ \frac{C_1}{2} & B & \frac{A_1}{2} & & & \\ & \frac{C_2}{2} & B & \frac{A_2}{2} & & \\ & & \ddots & \ddots & \ddots & \\ & & & \frac{C_{N-1}}{2} & B & \frac{A_{N-1}}{2} \\ & & & & \frac{(A_N+C_N)}{2} & B \end{bmatrix},$$

$\underline{\phi}$ is the column vector $\{\phi_0,\phi_1,\phi_2\ldots\phi_{N-1},\phi_N\}$ and $\underline{R}$ is the column vector $\{R_0,R_1,\ldots R_{N-1},R_N\}$. In combination with (11) it is necessary to determine X so as to satisfy the remaining boundary condition, in this instance $\phi = t$ at $\eta = 1$, so that X must satisfy

$$F(X) \equiv \frac{\phi_N(X)}{t} - 1 = 0 . \tag{12}$$

For a given estimate to the root of (12), equation (11) is solved for ϕ using the Thomas algorithm for a tridiagonal matrix. If X_1 and X_2 are two such estimates, and $F(X_1)$ and $F(X_2)$ are obtained from (12) using corresponding values of $\underline{\phi}$ obtained from the Thomas algorithm, then the method of "false position" gives

$$X_3 = \frac{X_1 F(X_2) - X_2 F(X_1)}{F(X_2) - F(X_1)} \tag{13}$$

as the next approximation to the root. The estimate corresponding to the larger of the two values $F(X_1)$ and $F(X_2)$ is discarded, and $F(X_3)$ determined from (12) after another application of the Thomas algorithm. The new estimate X_3 and the one retained from the original pair X_1 and X_2 are then used in (13) to obtain a further estimate X_4. This process is repeated until $F(X_n)$ is less than a certain specified quantity; X_n and $\underline{\phi}_n$ then represent the solution for the current value of the time variable. When advancing to a new time level, the initial estimate for X is taken to be the final value at the previous time level. In this way the calculation is advanced until the required time range has been covered.

It was found that the numerical results were not significantly affected by the reversal of the use of the two boundary conditions at $\eta = 1$; that is, with the first of (8) incorporated in the matrix equation and the second satisfied iteratively in the manner described. Further details and results are presented in [2].

REFERENCES

1. Crank, J., *Q. J. Mech. Appl. Math.* **10**, 220-231 (1957).
2. Ferriss, D. H. and Hill, Susan, National Physical Laboratory Report, NAC 45 (1974).
3. Landau, H. G., *Q. J. of Applied Maths.* **8**, 81-94 (1950).

The Nature of the Mushy Region in Stefan Problems with Joule Heating

E. Langham
(University of Quebec)

Mushy regions have been mentioned in several earlier papers (Tayler, p.129; Atthey, p. 183; Ockendon, p. 147). In particular they have been introduced in problems involving pure materials in order to avoid temperatures above the melting point in the solid phase when the principal cause of the phase change is internal (Joule) heating. Such temperatures are a natural consequence of the temperature gradient towards the phase boundary which is necessary to supply heat flow for the phase change.

It is of interest to consider in more physical detail what may happen near a phase boundary. The only restriction on the temperature of the solid phase is the requirement of equilibrium of the two phases at the phase boundary. This requirement is not *a priori* physically necessary elsewhere.

The stable phase is that which has the lowest free energy per unit mass (chemical potential). Now the initiation of a new phase takes place through the appearance, in the parent phase, of nuclei, which possess a surface free energy proportional to their surface area. Thus the free energy of a sufficiently small nucleus is always greater than that of the same mass of the parent phase and such a nucleus is unstable.

There is thus a barrier to the formation of a new phase which cannot be overcome unless the temperature exceeds the equilibrium temperature at the phase boundary by an amount sufficient for the combined volume and surface free energies of the first nucleus to be formed to fall below the free energy of the same mass of the parent phase. This excess temperature permits the existence of a temperature gradient towards the phase boundary. If the new phase is seeded away from the phase boundary, as the nucleus grows its equilibrium temperature falls towards that at the advancing phase boundary.

Thus a mushy region can only exist physically if there is sufficient impurity present or if there is some mechanism for spontaneous nucleation. The presence of Tyndall flowers on ice shows that for some systems there is a sufficient lack of seed centres for only a few lenses of the new phase to form. In such circumstances the mathematical formulation which precludes mushy regions predicts results which are physically acceptable, and the formulation could even be adapted to take account of the nucleation effect.

Part 111. Thermal Explosion Papers

Introduction

The three papers in Part III are all concerned with thermal explosions and pose a completely different type of heat transfer and moving boundary problem from the previous application papers and fall outside the scope of the analytical papers of Part II. Here the basic problem is to examine mechanisms whereby the area of the interface between a fluid and a material at a higher temperature than the vapourisation temperature of the fluid may be increased by several orders of magnitude. This physical effect is deduced from analysis of recorded thermal explosions. The paper by Board and Hall presents the results of filmed laboratory experiments where thermal explosions have been successfully triggered and they postulate mechanisms to explain their observations. Buchanan seeks to examine quantitatively a particular model based on the asymmetric collapse of vapour bubbles and also to establish the size of particles resulting from the break up of the high temperature material caused by instability of the boundary. The third paper by Bevir and Fielding treats in more detail the fluid flow problem of the collapse of a vapour bubble.

The papers deal with an area of immediate practical concern and one where there is scope for theoretical development. It is also an area, however, where the elaboration of theoretical models and analysis needs the closest ties with experiment.

THERMAL EXPLOSIONS AT MOLTEN TIN/WATER INTERFACES

S.J. Board and R.W. Hall

(CEGB Berkeley Nuclear Laboratories)

1. INTRODUCTION

Thermal explosions can occur in a wide range of size scales, extending from laboratory interactions of a few grams of material, through foundry and reactor accidents in the 10-100kg range, to submarine volcanic explosions such as Krakatoa (4×10^{10} tonnes) which is the largest terrestrial explosion known to man [9].

On a laboratory scale, thermal interactions are usually very inefficient (10^{-2}-10^{-3} of the thermodynamic yield). However, there is some evidence that on a larger scale such events are much more vigorous (*e.g.* Hess and Brondyke [5]) and high efficiencies are possible. For instance, an analysis of the Quebec foundry incident [7] indicates that the explosive yield approached the thermodynamic yield, and the Krakatoa event, involving a 200MT yield, also appears to have had a high efficiency (> 10%).

It seems unlikely that spontaneous initiation would occur throughout a large interaction region simultaneously, but a single coherent interaction is necessary to produce high efficiencies. Thus it is possible that a spatial propagation process may occur in such events, so that an interaction at one point rapidly initiates interaction near by. Such a mechanism is a fundamental characteristic of chemical explosions but has only recently been considered in the context of thermal explosions (*e.g.* Colgate and Sigurgeirsson [4]), possibly since most laboratory experiments have been on a small scale and so have not shown clear evidence for spatial propagation.

Thermal explosions may be understood either as due to the rapid release of energy stored as superheat in a vaporisable liquid, where energy transfer from the hot liquid occurs at a relatively slow rate for some time before the explosion, or alternatively as a rapid energy transfer and simultaneous release occurring because of an associated rapid mixing process (fragmentation explosion). In the case of large scale explosions due to superheat, self-propagating release may be explained by shock nucleation, so the problem is resolved into understanding how a significant volume of highly-superheated liquid can be formed. For superheat to be a viable explanation, nucleation must be suppressed, *i.e.*, liquid-liquid contact must occur; however, in most metal-water interactions, the contact temperature would be above the critical point. Moreover, the energy density of the explosive yield observed is very much greater than that possible from a superheat-limited explosion. Thus a fragmentation explanation appears to be necessary for at least some energetic events.

In the case of a fragmentation explosion, spatial propagation will occur if the material motions produced by the explosion result in further mixing elsewhere. Two possible mechanisms are considered here: firstly the explosively-driven collapse of a vapour blanket separating the materials [2] and [6], and secondly the mixing due to Taylor and Kelvin- Helmholtz instabilities arising from the motions imparted by the expanding interaction region [4]. This paper presents the results of experiments intended to investigate these possibilities.

2. PRESSURE DRIVEN BLANKET COLLAPSE

2.1 Experiment

To investigate the effect of rapid vapour-blanket collapse, about 50 grams of molten tin at 800°C was poured onto a shallow crucible under water at an external pressure of 100 torr (see Fig. 1). The vapour-blanketed tin drop was

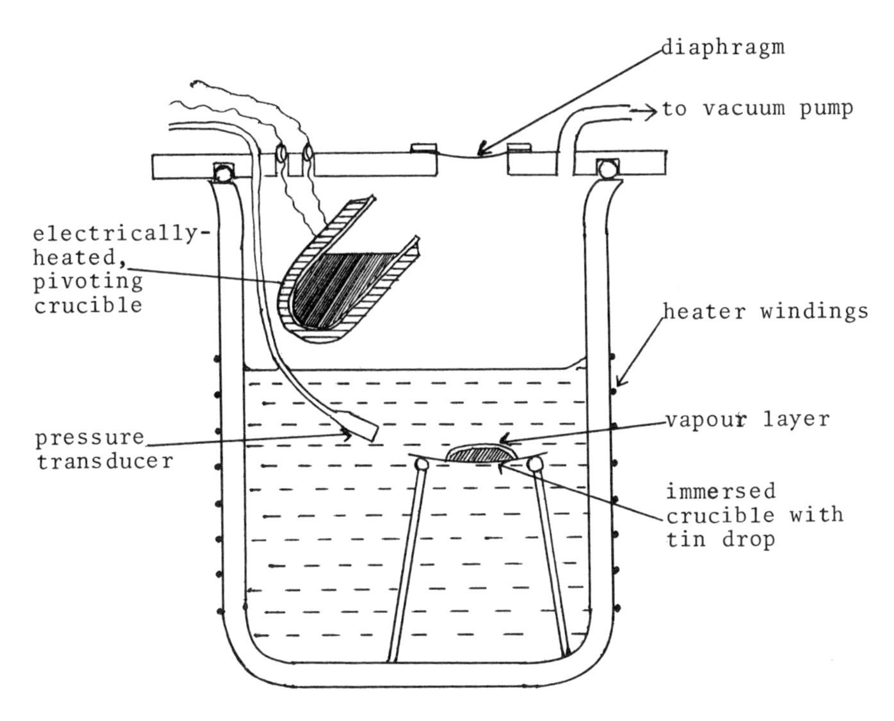

Fig. 1. Apparatus for testing the effect of a sudden pressure change on a stationary tin drop.

then subjected to a rapid increase in pressure ($\sim$1 bar) by rupturing a diaphragm connecting the apparatus to atmosphere. Pressure in the vessel was determined using an immersed Kistler 603B transducer, and events were photographed at $8000 ps^{-1}$. (This experiment differs from that described previously [2] in that the pressure rise was achieved much more rapidly, in 0.5ms instead of 20ms.)

The results showed that vapour-blanket collapse occurs within 1 ms of diaphragm rupture, and an explosion occurs within the following 250μs. A typical pressure trace (Fig. 2) shows an explosion pulse of $\sim$4 bar × 1ms, and no other events of significance either before or after the main pulse.

The experiment was repeated using aluminium which does not spontaneously explode under these conditions, and a similar, though less vigorous, explosion was obtained, thus confirming that the interaction was caused by the violent collapse of the blanket rather than being a coincidental result of achieving the conditions under which the metal normally explodes.

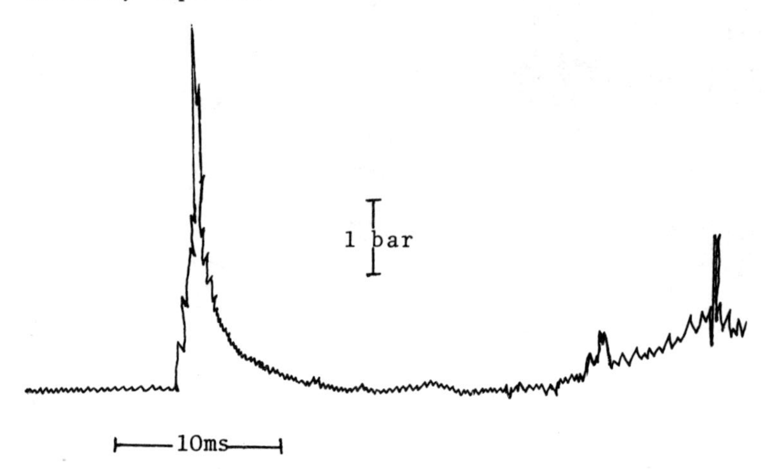

Fig. 2. Pressure record of a tin/water explosion produced on vapour blanket collapse in the apparatus shown in Fig. 1.

2.2 Discussion

The results of this experiment support the idea that explosion propagation can occur through pressure-driven blanket collapse: a local interaction increases the pressure in the surrounding liquid sufficiently to collapse the blanket on a larger area of adjacent material, and hence triggers its explosion.

Clearly, this mechanism can only operate if sufficient dynamic constraint is provided by the surrounding medium. In this experiment, however, the work done in collapsing the blanket (∿0.1J) is very much less than the yield of the resulting explosion (the kinetic energy in the water after the explosion is ∿10J, corresponding to a thermodynamic efficiency ∿0.2%) and so the energetics of the process are very favourable.

The energy yield of this experiment may be compared with that expected from a superheat-limited explosion, and also with that predicted on a fragmentation hypothesis. If the explosion were due solely to superheat around the undeformed drop, the maximum energy release could be estimated from the time available after blanket collapse for thermal-diffusive heating of the liquid boundary layer. In this experiment the time interval is very short ($\sim$0.25ms) so that the maximum release from superheat would be low ($\sim$1J). Since the experimental yield is $\sim$10J it would appear that superheat alone is not a sufficient explanation of this particular explosion, and fragmentation is required.

The fragmentation hypothesis of Board *et al* [2] (see also [3]) based on liquid jet penetration may be generalised to include the case of collapse of a vapour blanket in which there are irregularities such as vapour domes. The collapse of such domes in the blanket may be expected to lead to the formation of liquid jets which will penetrate the tin at velocities $\sim 10^4$ cm s^{-1} [8], giving rapid fine-scale mixing. In the present experiment, if we assume that the number of jets corresponds to the observed number of domes ($\sim$10) and the diameter of each jet to be $\sim\frac{1}{5}$ of the dome diameter then the total mass of water injected during collapse will be $\sim 3 \times 10^{-2}$g, giving a maximum potential yield $\sim$30 joules.

This mechanism could thus offer a sufficient explanation of the observed yield of 10J but not of explosions having efficiencies in excess of 1%. However, it is likely that further mixing of tin occurs during the explosive expansion; this is discussed later.

3. TROUGH EXPERIMENT

3.1 Experiment

The experiment described above indicated that explosion propagation through blanket collapse was likely on energy grounds. To test whether propagation would occur in practice, an experiment was performed using a trough-shaped crucible full of molten tin in which an interaction could be initiated at one end. About 200 grams of molten tin were poured into the aluminium crucible, 30cm long and of shallow "V" cross-section, immersed in an open tank of water at 80°C. The instrumentation consisted of 3 pressure transducers immersed in the water ∿10cm above the trough and spaced along its length, two thermocouples (∿100ms response) at the trough surface, and 5000ps^{-1} cinephotography.

At low tin temperatures (∿650°C) a spontaneous (and unexpected) interaction occurred near one end of the crucible. This was probably due to transition boiling in the still-wet region between the cooling tin and the newly-covered crucible surface. At higher initial temperatures (∿750°C) there was no spontaneous interaction and so an interaction was initiated by applying an impulse to one end of the crucible via a steel rod. (Initiation by localised increase of the subcooling by adding cold water proved unreliable.)

In general, whether the interaction was spontaneous or induced, the results showed that the initial event was weak and localised at one end of the trough and escalated in a single vapour-growth and collapse cycle to a vigorous but still localised interaction (as observed previously for single drops by Board *et al* [2]). During the growth phase of the "bubble" from this first vigorous interaction a second interaction occurred 10-15cm away, near the centre of the trough, and in turn, during the growth of this latter region a third interaction occurred at the far end.

The time interval between the successive interactions was typically in the range 2-5ms. The pressure pulses recorded from individual interactions were relatively low (0.7 bar) but of long duration (∿3ms). When subsequent interactions occurred within this pulse width, however, the pressures combined additively so that some resulting pulses were both higher (up to 1.5 bar) and longer (up to 10ms).

3.2 Discussion

These results provide some evidence of spatial propagation occurring in a thermal explosion, since a single interaction, either occurring spontaneously or externally induced, is observed to initiate a rapid sequence of interactions at other points. The sequence is not quite continuous, but the occasional additive pressure pulse suggests that continuous propagation could perhaps occur in other circumstances.

The mechanism providing the coupling between individual interactions is clearly different from the bubble growth and collapse process observed previously with single drops in cold water, since the new interaction occurs during the growth of the old bubble. Instead the behaviour is consistent with the results of the blanket collapse experiment discussed above, since the liquid displacement due to the growth of a bubble would be expected to cause the collapse of adjacent vapour blankets.

A possible explanation of the discontinuities in the propagation sequence could be that the relatively open geometry of the apparatus gives only weak coupling between neighbouring areas of tin; blanket collapse would thus occur only at regions of low stability. The stability of a vapour blanket apparently decreases rapidly with decreasing tin temperature (R.W. Hall, unpublished) and indeed the sites of the successive interactions were those where the tin would be coolest - at the ends of the trough and also near the centre where a metal guide tube was in contact with the tin.

4. THIN TANK EXPERIMENT

4.1 Experiment

To make the dynamic constraint comparable to that experienced by a full-scale explosion occurring at the horizontal interface between large regions of molten metal and water, a tank was constructed which could be regarded as containing a thin vertical slice of the full three-dimensional explosion. The open-topped tank, 20cm long × 3cm wide × 15cm high, with one face of perspex, was filled with water at 80°C. About 200 grams of molten tin at 700°C were poured into the tank and an interaction was initiated by tapping the base near one end. The high-speed film record, taken together with the pressure trace (Fig. 3) showed that the initial perturbation (*A*, Fig. 3) triggered only a very minor interaction (*B*) confined to the left-hand end of the tank. About 10ms after this however, a vigorous explosion began (*C*) and propagated rapidly along the tin/water interface (see photographs in Fig. 4), travelling 15cm along the tank in 3ms before the pressure dropped (*D*). The apparent propagation velocity was $\sim 5.10^3$cm s^{-1}, whilst the vertical expansion velocity was about half of this. As the pressure dropped, the interaction front slowed down to the vertical expansion velocity. Normal bubble expansion then continued, and after a delay of 5ms the remaining tin at the right-hand end of the vessel interacted (*E*).

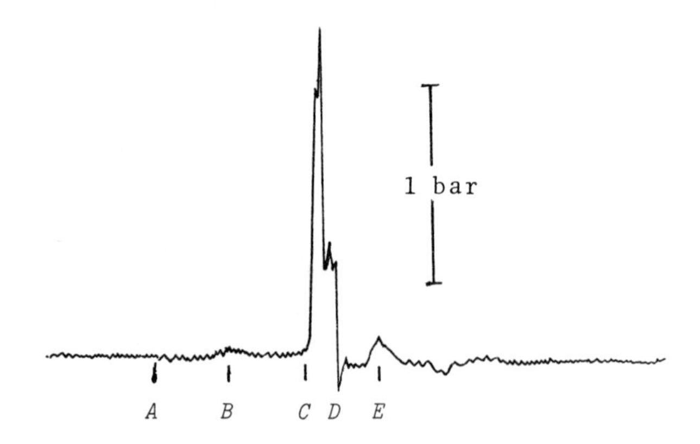

Fig. 3. Pressure record of an event initiated at one end of a long tank.

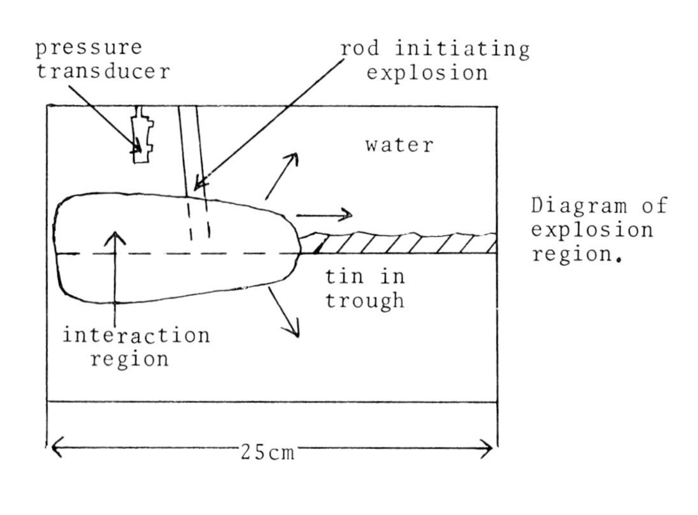

Diagram of explosion region.

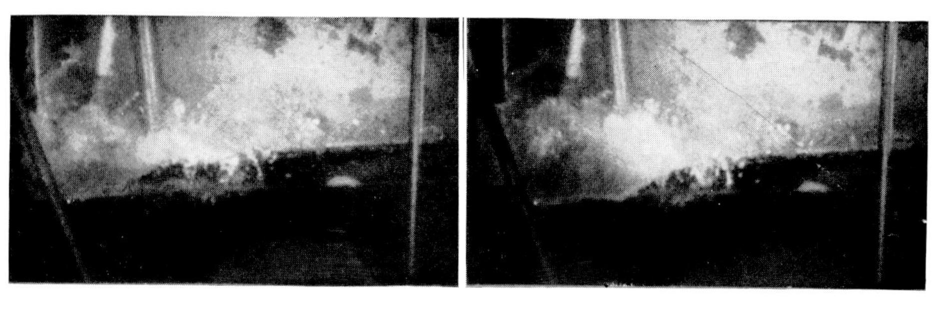

0ms 0.8ms

1.6ms 2.4ms

Fig. 4. Propagation of explosion "thin tank" geometry.

4.2 Discussion

This experiment showed evidence for a much more continuous propagation process than that observed in the trough experiments. Whilst this could be due to more uniform conditions in the tin it seems more likely that the increase in dynamic constraint provided by the thin vessel was the important factor.

The preferential growth of the bubble along the tin/water boundary suggests that pressure close to the interaction front is higher than that elsewhere in the bubble. This suggests that vapour production, and thus the mixing which produces it, must be occurring largely in the region of the front. Since the interaction region contains liquids (water and tin) as well as vapour, the local velocity of sound will be low, and at the observed propagation velocity (5×10^3 cm s^{-1}), compressible flow or shock-wave phenomena may be important in sustaining the required non-uniformity of pressure in the bubble.

The rapid motion of the front also suggests that mixing due to Taylor and Kelvin-Helmholtz instabilities could occur in this region. If the degree of mixing is limited by surface tension (σ) rather than viscosity, a characteristic mixing dimension, $\frac{6\sigma}{\eta\rho u^2}$ may be defined, where η represents the efficiency with which kinetic energy $\frac{\rho u^2}{2}$ is transformed into surface energy. For a propagation velocity $\sim 5 \times 10^3$ cm sec^{-1} and mixing efficiencies of about 10%, particles of a size $\sim 10\mu$ would be generated; these, if blanketed by a vapour layer $\sim 10^{-3}$ cm [1] could easily maintain the pressurisation of the expanding bubble. Thus if the mixing efficiency is above 10% a significant contribution to the energy released could result from mixing caused by the motion of the front.

In the blanket collapse experiment, where the expansion velocity is $\sim 10^3$ cm s^{-1}, the equivalent required mixing efficiency is $\sim 100\%$ so that a significant yield from explosion mixing is less likely.

5. CONCLUSIONS

The results of several small-scale experiments demonstrate that spatial propagation can occur in thermal explosions and also suggest that the observed explosions are primarily due to fragmentation rather than superheat. It has been shown that propagating fragmentation or mixing can occur through self-driven vapour blanket collapse. In a continuously propagating explosion however, the Taylor and Kelvin-Helmholtz instabilities arising from the motion of the interaction region probably become the dominant mixing processes.

There is some evidence that the flow within the two-phase interaction region is compressible.

REFERENCES

1. Board, S.J., Duffey, R.B., Farmer, C.L. and Poole, D.H., *Nuclear Sci. Eng.* **52**, 433-438 (1973).
2. Board, S.J., Farmer, C.L. and Poole, D.H., CEGB Report RD/B/N 2423 (1972).
3. Buchanan, D.J. and Dullforce, T.A., *Nature*, **245**, 32-34 (1973).
4. Colgate, S.A. and Sigurgeirsson, T., *Nature*, **224**, 552-555 (1973).
5. Hess, P.D. and Brondyke, K.J., *Met. Prog.*, 93-100 (1969).
6. Jakeman, D. and Potter, R., "Fuel Coolant Interactions," *Int. Conf. Eng.* Fast Reactor for Safe and Reliable Operation, Karlsruhe (1972).
7. Lipsett, S.G., *Fire Technology*, **2**, 118-126 (1966).
8. Plesset, M.S. and Chapman, R.B., *J. Fluid Mech.* **47**, 283-290 (1971).
9. Symons, G., Report of the Krakatoa Committee of the Royal Society (1888).

FUEL-COOLANT INTERACTION THEORY

D.J. Buchanan
(UKAEA, Culham Laboratory)*

1. INTRODUCTION

It is well known that when a hot liquid comes into contact with a cold vaporisable liquid an explosion of considerable violence may occur. This type of explosion is called alternatively a vapour explosion, a thermal explosion or fuel-coolant interaction (FCI), the hot liquid being the fuel and the cold liquid the coolant. These interactions are not the result of chemical change; the energy source is the excess heat in the fuel. FCIs have been observed between cold water and various molten metals (e.g. tin, indium, steel and aluminium), in the nuclear field between molten uranium dioxide (the fuel) and liquid sodium (the coolant), and in the chemical industry between liquefied natural gas (the coolant) and water (the fuel). Witte *et al* [49] quote many examples of FCIs. The term "FCI" has, in the past, been used only of interactions between reactor fuel and reactor coolant but we use it generically to cover all interactions. The interaction between UO_2 and Na has occurred only in very specialised circumstances and there is much interest in determining whether FCIs are physically possible in Na/UO_2 reactors.

In Section 2 it is shown that an exact theoretical description of FCIs based on established physical equations is unlikely, and that a model approach must be used. After referring to previous theoretical models, a brief description of one particular model is given. In Section 3 some moving boundary problems associated with this model are discussed and

---

* Present address: Mining Research and Development Establishment National Coal Board, Burton-on-Trent.

in Section 4 we show how a set of equations may be obtained which may be used to calculate part of the particle size distribution function that results from an FCI. No numerical results are given; in the case of the model discussed in Section 3 results have been given elsewhere, and calculations on the particle size distribution function have not yet been completed.

2. FORMULATION OF FCI THEORY

Chemical analysis of the debris resulting from an FCI shows that little if any chemical change occurs [18], and Lipsett [31] has concluded that the excess heat in the fuel is the energy source. Lipsett [31] has also shown that to transfer sufficient heat within the timescale of the explosion is impossible by any known means unless there is a large area of contact between the fuel and the coolant. Witte *et al* [49] have stated that heat transfer rates of order 10^{10} BTU/Hr-ft^2 are required, and since heat transfer rates during boiling processes rarely exceed 10^7 these authors also conclude that a large increase in surface area, $\sim 10^3$, is a necessary criterion for an FCI. K.V. Roberts (private communication), by considering the heat diffusion problem in the fuel, has also shown that an area increase of order 10^3 is required. Thus one of the basic physical problems in FCI theory is to show how this rapid area increase occurs.

If we assume that the only thermodynamic phases present in the whole system are liquid and vapour coolant, and liquid and solid fuel, then a complete theory involves a maximum of six moving boundaries. Of course, the presence of impurities complicates matters even more but we shall assume that the system consists only of pure fuel and coolant. Each of the fluid regions is governed by continuity and energy equations and a Navier-Stokes equation, and the solid region is described by the heat conduction equation. This system of equations is difficult to solve for fixed boundaries but in this instance the occurrence of moving boundaries makes the solution even more intractable. A numerical *ab initio* solution is also quite impracticable. Space scales may range from 1m to 10^{-5}m, the size of the fine-scale debris produced by the interaction itself, and to allow 10^5 mesh points in each direction of a three-dimensional mesh and 10^5 timesteps, at a calculation speed of 1μsec/point, would require 300 years for a complete calculation. Thus a theory of FCI based on an established set of fundamental equations is inconceivable at present. In fact it is not certain that the fundamental set of equations discussed above are a sufficient set of equations for an FCI. Experiment shows that FCIs are sensitive to a large number of parameters which are not included in these equations.

In this situation of impracticability we use physical intuition to break the problem into a number of separate parts each of which is solved individually, and then the solutions are synthesised to give the solution, or model solution, to the original problem. This technique has been adopted in all theoretical discussions of FCIs to date (at least to the author's knowledge) and this paper is no exception.

The central problem is, of course, to devise a mechanism whereby the contact area between fuel and coolant is rapidly increased. Several models have been proposed and some have been discounted. See, for example, the papers of Bankoff and Fauske [2], Board *et al* [6], Brauer *et al* [7], Buchanan [8], Buchanan and Dullforce [10], [11], Caldarola and Kastenberg [13], Cronenberg *et al* [16], Enger *et al* [17], Fauske [19], [20], Groenveld [22], Hesson and Ivins [24], Ivins [27], Katz [29], Katz and Sliepcevich [30], Long [32], Nelson [34], Nelson and Kennedy [35], Potter and Jakeman [39], [40], Roberts [42], Sallack [43], Schins [44], Swift [45], Swift and Pavlik [46], Witte *et al* [49], Yang [50] and Zyszkowski [51] and [52]. This list also gives some idea of the variety of situations in which FCIs may occur. Some of the models discussed by the above authors are qualitative descriptions and make no attempt at quantitative prediction. However, it is not sufficient to say that a certain scheme results in area increase; it must also be shown that the area increase is large enough and sufficiently rapid to get an adequate amount of heat out of the fuel within the timescale of the explosion. A second type of empirical model also exists in the literature. See, for example, the papers of Biasi *et al* [5], Caldarola [12], Cho *et al* [14], Jacobs and Thurnay [28], Morgan [33], Randles [41], Syrmalenios [47] and Tezuka *et al* [48]. These models assume a certain area increase and then calculate the resulting pressure in reactor geometry. By adjusting a large number of parameters the predicted pressure-time histories can be compared with experiment. Since the reason for the area increase is one of our main concerns, these models will not be mentioned further here.

For the remainder of this paper we shall concentrate on one of the models of the first group. Experiments by Board *et al* [6] suggest that the following model may be appropriate. The justification for the model is discussed in that paper as well as those of Buchanan [9] and Buchanan and Dullforce [10] and [11]. The interaction is divided into five stages, the last four of which occur cyclically.

Stage 1: as a result of some triggering mechanism (for example, the onset of transition boiling as the hot liquid cools) the liquids come into intimate contact and a vapour bubble is formed.

Stage 2: the vapour bubble expands and then collapses as a result of condensation in the subcooled coolant. Because of the density and/or viscosity difference between fuel and coolant (and also possibly because a solid crust has formed on the fuel) the bubble collapse is asymmetric and a high velocity jet of liquid coolant, directed towards the fuel, is formed.

Stage 3: the jet of coolant penetrates the fuel and, since it has a high velocity, its subsequent disintegration and hence increase in fuel-coolant contact area is extremely rapid.

Stage 4: as the jet of coolant penetrates and breaks up, heat is transferred from the surrounding fuel to the jet and since the surface area is increasing rapidly the total heating rate also increases rapidly.

Stage 5: when the jet has been heated to a certain temperature it suddenly vaporises and a high pressure vapour bubble forms. The bubble expands into the subcooled coolant and the process advances from stage 2 again.

Because of the cyclic scheme of the mechanism some of the energy released from the fuel during each cycle is fed into subsequent cycles and it is this feedback that enables a small initial perturbation to grow. Note that the model provides a mechanism for increasing the contact area between fuel and coolant which is essential to explain the observed rapid heat transfer.

Preliminary calculations on this model have been performed by Buchanan [9] and a brief description of some of the results was given by Buchanan and Dullforce [10]. Peckover *et al* [36] have used the model to describe the interaction of magma from an underwater volcano with the surrounding sea, and Ashby *et al* [1] recognised that the model described a fluid instability of more general application than had hitherto been supposed and applied the results to the interaction of matter with antimatter.

Stage 1 is solely a means of supplying the initial perturbation which causes the first bubble to form adjacent to the fuel surface, and it is thought that the perturbation is the result of the collapse of the vapour layer between the fuel surface and the liquid coolant. Although this in itself is an interesting moving boundary problem it will not concern us here.

3. MOVING BOUNDARY PROBLEMS ASSOCIATED WITH STAGES 2-5

Stage 2

The initial condition for the ith cycle is a spherical bubble, initial pressure P_i and radius R_i, which expands in the surrounding incompressible coolant of density ρ_ℓ and ambient pressure P_0. To date calculations have been done on the assumption that no heat transfer or condensation occurs during the expansion phase until the maximum radius is reached. Clearly this is an idealised situation and over-estimates the maximum size of the bubble. A more realistic treatment which includes heat transfer and condensation leads to the equations (M.K. Bevir and R.J. Loader (private communication))

$$R\ddot{R} + \tfrac{3}{2}\dot{R}^2 = \frac{P_w(T_w) - P_0}{\rho_\ell}\,, \tag{3.1}$$

$$T_w(t) - T_0 = -\left(\frac{\kappa_\ell}{\pi}\right)^{\frac{1}{2}} \frac{L_v}{4\pi k_\ell} \int_0^t \frac{\frac{dm}{d\tau}d\tau}{\left[\int_\tau^t R^4(\tau')d\tau'\right]^{\frac{1}{2}}}\,, \tag{3.2}$$

$$\dot{m} = \rho_v\left(4\pi R^2\dot{R} + \tfrac{4}{3}\,\frac{\pi R^3}{\gamma P}\,\dot{P}_w\right)\,, \tag{3.3}$$

$$P_w = f(T_w)\,. \tag{3.4}$$

The density of the vapour is given either by the vapour equation of state or else the adiabatic compression law.

These equations are an extension of the model of Florschuetz and Chao [21]. (3.1) is the familiar Rayleigh equation, (3.2) is the zeroth order solution of Plesset and Zwick [37] for the heat diffusion across a spherical boundary with radial motion, (3.3) takes account of changes of phase and (3.4) is the equation of state. R is the radius of the bubble at time t, P_w is the pressure in the bubble, T_0 and T_w

are the temperatures of the bulk coolant and the bubble wall respectively, $\kappa_\ell = k_\ell/\rho_\ell c_\ell$ and L_v are the thermal diffusivity and latent heat of the coolant, ρ_v and γ are the density and adiabatic constant of the vapour. M.K. Bevir (Culham Laboratory) and R.J. Loader (Reading University) are developing a computer code to solve (3.1) to (3.4) and when completed it will take the place of the simple treatment assumed in previous calculations. Future developments will allow for non-condensible gas within the bubble.

The second part of the stage 2 calculation involves the collapse of the bubble. It is well known that when an initially spherical cavity collapses adjacent to a solid wall the collapse is axisymmetric and a jet of liquid, directed towards the wall, is formed. Plesset and Chapman [38] have done a numerical simulation of this situation and they find that the final jet velocity scales as $(\Delta P/\rho_\ell)^{\frac{1}{2}}$ where ΔP ($= P_0$ in this case) is the pressure difference causing the collapse. The final jet dimensions scale as the initial (pre-collapse) cavity radius. Thus the jet dimensions are determined by R_m and consequently by R_i and P_i. The velocity, length and diameter of the jet when it strikes the fuel surface are

$$\left.\begin{aligned} V_0 &= V_c\left(\frac{\Delta P}{\rho_\ell}\right)^{\frac{1}{2}} \\ L_0 &= L_c R_m \\ d_0 &= d_c R_m \;. \end{aligned}\right\} \qquad (3.5)$$

The length of the jet, L_0, is defined to be the diameter of the bubble along the axis of the jet as it strikes the wall (see Plesset and Chapman [38], Fig. 1). The jetting that occurs here is a result of the lack of spherical symmetry in the local surroundings and thus the constants V_c, L_c and d_c are determined by the degree of departure from spherical symmetry. For bubble collapse adjacent to a solid wall

$$\left.\begin{aligned} V_c &= 13 \\ L_c &= 0.493 \\ d_c &= 0.237. \end{aligned}\right\} \qquad (3.6)$$

If the adjacent layer of fuel is liquid then the collapse will still be axisymmetric (due to the density difference and/or viscosity difference) but the numerical values of the constants may be different. In addition, the values of the constants will also be different if the bubble is not tangential to the fuel surface. Bevir and Fielding (p.286) discuss bubble collapse in more detail. A recent paper by Guy and Ledwidge [23] takes into account mass transfer as the bubble collapses.

Stage 3

The jet of coolant enters the fuel with velocity V_0. As the jet penetrates it disintegrates and mixes with the surrounding fluid. J.P. Christiansen and J.A. Reynolds (private communication) (see also [15]) have performed a numerical simulation of the mixing process. They show that the length of the jet (the calculation is two-dimensional) increases exponentially with a time constant proportional to d_0/V_0. This length increase represents the mixing. By using Batchelor's [3] results to relate length increase to area increase, we represent the increase, by mixing, of the contact area between the fuel and the liquid jet by

$$A = A_0 e^{t/\tau} \tag{3.7}$$

$$\tau = \frac{11}{4}\left(\frac{\rho}{\rho_\ell}\right)^{\frac{1}{2}} \frac{d_0}{V_0} \tag{3.8}$$

where ρ is the density of the fuel and A_0 the initial contact area. Further details are given in Buchanan [9]. Whilst the jet remains liquid its volume is almost constant. Consequently, since the area increases exponentially, the average distance, s, between material surfaces must decrease exponentially

$$s = s_0 e^{-t/\tau} \; . \tag{3.9}$$

There are three factors limiting the area increase. Firstly, if the fuel surrounding the jet solidifies then no further increase in contact area can occur. Secondly, if the jet vaporises then the conditions are radically altered. A high pressure bubble now exists and this blows out the surrounding fuel. Thirdly, if the disintegration process is that envisaged by Hinze [25], [26], then as the average distance between surfaces decreases the Weber number decreases and is eventually equal to the critical Weber number below which further disintegration cannot occur. We denote the minimum value of s by s_m regardless of the manner in which s_m is attained.

Stage 4

As the jet penetrates the fuel, heat transfer occurs. Fig. 1 represents a cross-sectional view of the situation sometime after the initial impact. We assume that heat transfer takes place one dimensionally across each element of fuel-coolant-fuel as shown in Figs. 1 and 2.

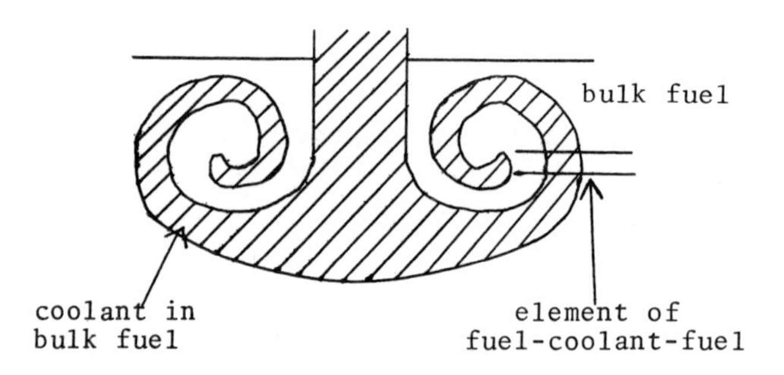

Fig. 1. Schematic cross-section of a jet of coolant sometime after penetration.

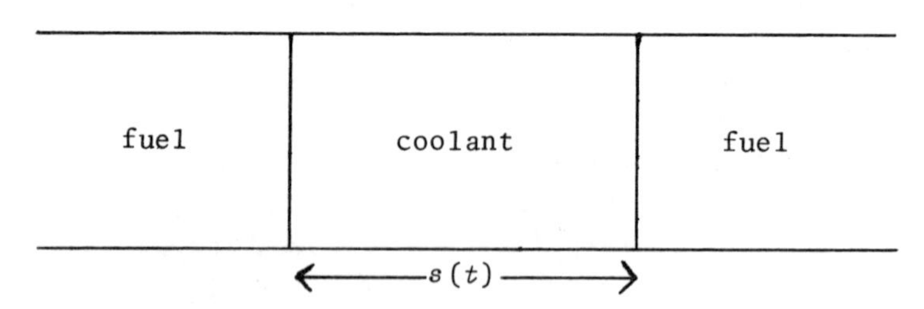

Fig. 2. Element of fuel-coolant-fuel in which heat transfer occurs one dimensionally.

The thickness of the element of the coolant, $s(t)$, is a decreasing function of time until the minimum size is reached. Thereafter, the thickness is constant. The heat transfer problem has been treated approximately [8], [9] by assuming that the temperature and pressure are constant within some region of the coolant. The temperature in the coolant is then given by

$$c_{\ell}\rho_{\ell}x\dot{T} = r_1 - r_2T \tag{3.10}$$

where r_1 is a parameter specifying the rate of heating of the jet per unit contact area due to heat transfer from the surrounding hot fluid. The term r_2T is inserted to take account of the fact that the heat transfer rate per unit contact area decreases as the temperature of the jet increases. x is the length of the coolant region being heated and takes the following functional forms in turn.

Initially, the region of coolant effectively heated is of length $2\sqrt{\kappa_{\ell}t}$ at both ends of the slab of fuel-coolant-fuel. Thus $x = 4\sqrt{\kappa_{\ell}t}$. When $4\sqrt{\kappa_{\ell}t}$ is equal to the thickness of the element, s, the coolant is heated throughout and the value of x is given by (3.9). Eventually $s(t)$ is equal to the minimum value and in this case $x = s_m$. (3.10) is easily solved in each case.

A better approach would be to solve the equation

$$\frac{\partial T}{\partial t} + V_x\frac{\partial T}{\partial x} = \kappa_{\ell}\frac{\partial^2 T}{\partial x^2} \tag{3.11}$$

within the region $-\frac{1}{2}s(t) \leqslant x \leqslant \frac{1}{2}s(t)$ with $s(t)$ given by (3.9) together with the cut-off s_m. V_x is the x-component of the velocity.

Stage 5

This stage encompasses the vaporisation of the jet and the resulting formation of a high-pressure, high-density gas bubble. An essential feature of stage 5 is the link back to stage 2 by identifying the vaporised jet with the bubble for the next cycle. The subsequent motion of the gas bubble is described by stage 2. Although moving boundaries do exist within stage 5 their effect has not been taken into account. In particular, it is assumed that the finite part of the jet which vaporises does so instantaneously thus giving a large initial pressure P_i. In reality the vaporisation process takes a finite time, perhaps 10^{-6}sec, and thus a more exact calculation of P_i would take into account the expansion that could occur within that time. However, this has not been done to date. For further details Buchanan [9] should be consulted.

The calculations outlined in this section, although somewhat crude, enable the consequences of the proposed mechanism to be examined. A full description of some illustrative calculations is contained in Buchanan [9]. One of the principal results obtained therein is that the strength of the interaction is reduced as the external pressure is increased and can be inhibited entirely if the pressure is large enough. The numerical value of this limiting pressure is of course dependent on the detailed calculations and for that reason, if no other, some of the improvements suggested in this paper will have to be incorporated before detailed comparison with experiment can be made.

4. PARTICLE SIZE ESTIMATION

The model presented in the previous section contains no attempt to calculate the particle size. Indeed the only mention of particle size is through the parameter s_m, and calculations [9] show that the results are not very sensitive to quite large ranges of this parameter. There are two possible sources of the fine scale debris.

1. As the jet of coolant penetrates and mixes with the hot fuel, small particles of fuel must result as the heat is transferred to the coolant. These particles are parametrised by s_m.

2. When the coolant vaporises and expands rapidly, the surrounding fuel in the locality of the vapour is broken into large blobs. If the velocity of the blobs is large enough, further disintegration will occur by the process envisaged by Hinze [25], [26].

The first process will not concern us further. We now indicate how the second mechanism may be described in an idealised manner.

We postulate that a drop breaks into two equal spherical drops if, at its maximum deviation from sphericity, the Weber number is greater than some critical value We_c. We further suppose that the drops will continue to split in this manner until the surface temperature drops to the freezing point of the liquid fuel, T_m. Thus the eventual particle size may be calculated if we can find:

(*i*) the time, t_s, taken for a drop of given size and initial velocity to split;

(*ii*) the time, t_m, taken for the temperature at some point on the surface of the drop to fall to T_m.

The most helpful simplification in either of these calculations results from assuming that the drop only deviates slightly from being spherical. Then the stress distribution at its surface can be written down explicitly for potential flow in the surrounding coolant. The velocity of the drop may then be calculated from its equation of motion, and finally the linearised deviation of the drop from spherical can be written down following Hinze for either slow or potential flow in the drop.

The computation of t_m involves solving the relevant heat conduction and convection equations in the drop of fuel and surrounding coolant, given the velocity fields used in the disintegration calculation. Even assuming a nearly spherical drop, the calculation is complicated by the phase changes which occur in the layer of coolant vapour which forms around the surface. Almost certainly a numerical solution of this problem is required, although it may be that the case of a thin vapour layer is amenable to a perturbation analysis.

If these calculations can be performed successfully the computation procedure is as follows:-

(*a*) determine the time t_s at which the maximum deformation is reached;

(*b*) determine t_m;

(*c*) calculate the Weber number, We_s, at time t_s.

Then

(*i*) if $t_m < t_s$, or $t_m > t_s$ and $We_s < We_c$ no more disintegration occurs;

(*ii*) if $t_m > t_s$ and $We_s > We_c$ splitting occurs;

(*iii*) if disintegration has occurred, the temperature and velocity fields, which give the initial conditions for the next calculation, must be determined. The calculation then continues from (*a*) above.

If this process occurs i times to an initial single particle of radius R_0 then 2^i particles result each with radius $R_0/2^{i/3}$ and the eventual particle size distribution is thus simply determined by the initial distribution.

ACKNOWLEDGEMENTS

I am grateful to Dr. M.K. Bevir and the editors for their helpful comments.

REFERENCES

1. Ashby, D.E.T.F., Buchanan, D.J. and Peckover, R.S., *Nature*, **247**, 272 (1974).

2. Bankoff, S.G. and Fauske, H.K., *Int. J. Heat Mass Trans.*, **17**, 461 (1973).

3. Batchelor, G.K., *Proc. Roy. Soc.* **A213**, 349 (1952).

4. Bevir, M.K. and Fielding, P.J., this conference.

5. Biasi, L., Clerici, G.C., Cecco, L. and Scarno, G., paper presented to the Second Specialist Meeting on Sodium Fuel Interaction in Fast Reactors, Ispra, Italy (1973).

6. Board, S.J., Farmer, C.L. and Poole, D.H., CEGB report no. RD/B/N2423 (1972).

7. Brauer, F.E., Green, N.W. and Mesler, R.B., *Nuc. Sci. Eng.* **31**, 551 (1968).

8. Buchanan, D.J., *High Temps-High Press*, **5**, 531 (1973).

9. Buchanan, D.J., to appear in *J. Phys. D: Appl. Phys.* (1974).

10. Buchanan, D.J. and Dullforce, T.A., *Nature*, **245**, 32 (1973).

11. Buchanan, D.J. and Dullforce, T.A., paper presented to the Second Specialist Meeting on Sodium Fuel Interaction in Fast Reactors, Ispra, Italy (1973).

12. Caldarola, L., *Nuc. Eng. Design*, **22**, 175 (1972).

13. Caldarola, L. and Kastenberg, W.E., paper presented to the Second International Conference on Structural Mechanics in Reactor Technology, Berlin (1973).

14. Cho, D.H., Ivins, R.O. and Wright, R.W., Proceedings of Conference on New Developments in Reactor Mathematics and Applications, USAEC Report CONF-710302, 25 (1971).

15. Christiansen, J.P., *J. Comp. Phys.* **13**, 363 (1973).

16. Cronenberg, A.W., Chawla, T.C. and Fauske, H.K., *Trans. Am. Nuc. Soc.*, **17**, 351 (1973).

17. Enger, T., Hartman, D.E. and Seymour, E.V., paper presented at the Cryogenic Engineering Conference, Boulder, Colorado, (1972).

18. Epstein, L.F., "Metal-Water Reactions: VII Reactor Safety Aspects of Metal-Water Reactions," US Atomic Energy Commission Publication GEAP 3335 (1960).

19. Fauske, H.K., *Nuc. Sci. Eng.* **51**, 95 (1973).

20. Fauske, H.K., paper presented at the Second Specialist Meeting on Sodium Fuel Interaction in Fast Reactors, Ispra, Italy (1973).

21. Florschuetz, L.W. and Chao, B.T., *Trans. ASME J. Heat Trans.*, **C87**, 209 (1965).

22. Groenveld, P., *Trans. ASME J. Heat Trans.* **C94**, 236 (1972).

23. Guy, T.B. and Ledwidge, T.J., *Int. J. Heat Mass Trans.* **16**, 2393 (1973).

24. Hesson, J.C. and Ivins, R.O., ANL Chemical Engineering Division Semi-Annual Report ANL-7425, July-December, 159 (1967).

25. Hinze, J.O., *App. Sci. Res.* **A1**, 263 (1948).

26. Hinze, J.O., *App. Sci. Res.* **A1**, 273 (1948).

27. Ivins, R.O., ANL Chemical Engineering Division Semi-Annual Report ANL-7399, November, 162 (1967).

28. Jacobs, H. and Thurnay, K., Proceedings of Conference on Engineering of Fast Reactors for Safe and Reliable Operation, Karlsruhe, 936 (1972).

29. Katz, D.L., *Chem. Eng. Prog.* **68**, 68 (1972).

30. Katz, D.L. and Sliepcevich, C.M., *Hydrocarb. Proc. Petrol. Refin.* November, 240 (1971).

31. Lipsett, S.G., *Fire Technology*, **2**, 118 (1966).

32. Long, G., *Metal Progress*, **71**, 107 (1957).

33. Morgan, K., AWRE Aldermaston TRG Report 2459 (R/X) (1973).

34. Nelson, W., paper presented at the Black Liquor Recovery Advisory Committee Meeting, Atlanta (1972).

35. Nelson, W. and Kennedy, E.H., *Paper Trade Journal*, **140**, 50 (1956).

36. Peckover, R.S., Buchanan, D.J. and Ashby, D.E.T.F., *Nature*, **245**, 307 (1973).

37. Plesset, M.S. and Zwick, S.A., *J. App. Phys.* **23**, 95 (1952).

38. Plesset, M.S. and Chapman, R.B., *J. Fluid Mech.*, **47**, 283 (1971).

39. Potter, R. and Jakeman, D., paper presented at the CREST meeting on Sodium-Fuel Interaction in Fast Reactor Safety, Grenoble (1972).

40. Potter, R. and Jakeman, D., paper presented to the International Conference on Engineering of Fast Reactors for Safe and Reliable Operation, Karlsruhe (1972).

41. Randles, J., Euratom report EUR 4592e (1971).

42. Roberts, K.V., paper presented at the CREST meeting on Sodium-Fuel Interaction in Fast Reactor Safety, Grenoble (1972).

43. Sallack, J.A., *Pulp and Paper Magazine of Canada*, **56**, 114 (1955).

44. Schins, H., paper presented at the Second Specialist Meeting on Sodium-Fuel Interaction in Fast Reactors, Ispra, Italy (1973).

45. Swift, D., ANL Chemical Engineering Division Semi-Annual Report ANL-7125, July-December, 192 (1965).

46. Swift, D. and Pavlik, J., ANL Chemical Engineering Division Semi-Annual Report, ANL-7135, 187 (1966).

47. Syrmalenios, P., Centre d'Etudes Nucléaires de Grenoble, Report CEA-R-4432 (1973).

48. Tezuka, M., Suzuki, K., Sasanuma, K. and Nagasima, K., paper presented at the Second Specialist Meeting on Sodium-Fuel Interaction in Fast Reactors, Ispra, Italy (1973).

49. Witte, L.C., Cox, J.E. and Bouvier, J.E., *J. Metals*, **22**, 39 (1970).

50. Yang, K., *Nature*, **243**, 221 (1973).

51. Zyszkowski, W., paper presented to the International Meeting on Reactor Heat Transfer, Karlsruhe (1973).

52. Zyszkowski, W., paper presented to the Second Specialist Meeting on Sodium-Fuel Interaction in Fast Reactors, Ispra, Italy (1973).

Numerical Solution of Incompressible Bubble Collapse with Jetting

M.K. Bevir and P.J. Fielding
(UKAEA, Culham Laboratory)

It is known, experimentally and theoretically, that jets of liquid can occur inside a bubble or cavity collapsing under the imposition of a pressure far away. Apart from the unstable motions of the interface, the formation of jets requires some departure from spherical symmetry in the collapse conditions. This spherical asymmetry might be caused by the presence of a flat wall, a gravitational field or some other effect. Such jets move considerably faster than the rest of the collapsing cavity and are thought to cause cavitation damage in cases where the cavity touches the wall. They may also be a source of intense sound production, or increase the heat transfer in fuel coolant interactions (Board, p.259; Buchanan, p.270).

The aim of our work has been the production of a numerical code sufficiently fast to investigate the formation of jets systematically. The formulation, which can include surface tension at the cavity interface, also allows the investigation of the growth of cavities. It could thus be used in certain circumstances to describe the growth of bubbles at a heated interface provided the growth was determined by the introduction of vapour into the bubble at a rate specified by some other mechanism. At present the effect of heat conduction within the liquid is not included, though this might be added at a later stage using the thin thermal boundary layer approximation.

The problem thus envisaged is the time dependent motion of a cavity of an incompressible liquid in an axisymmetric geometrical situation. The pressure far from the cavity,

$p_\infty(t)$, and the internal pressure, $p(t)$, are prescribed. Alternatively the internal pressure may be determined from the volume of the cavity and some gas law or other assumption.

The mathematical formulation of the problem, (*e.g.* Plesset and Chapman [2]), is simple. The motion $\underset{\sim}{v}$ of the incompressible liquid starting from rest and ignoring viscosity is determined by a potential function ϕ satisfying

$$\nabla^2\phi = 0. \tag{1}$$

The boundary conditions on ϕ are as follows. Far from the bubble $\underset{\sim}{v} = 0$ so that we may take $\phi = 0$. On a wall $v_n = \frac{\partial\phi}{\partial n} = 0$. It remains only to write down the boundary conditions at the cavity interface. Bernoulli's equation in unsteady motion gives

$$\frac{\partial\phi}{\partial t} + \frac{v^2}{2} = \frac{p_\infty - p}{\rho} - gz + \text{surface tension terms}$$

where p is the pressure in the cavity and g is the gravitational **field.** Thus, defining the convective derivative

$$\frac{D\phi}{Dt} = \frac{\partial\phi}{\partial t} + (\underset{\sim}{v}.\nabla)\phi$$

$$= \frac{\partial\phi}{\partial t} + v^2,$$

we obtain

$$\frac{D\phi}{Dt} = \frac{p_\infty - p}{\rho} + \frac{v^2}{2} - gz + \text{surface tension terms.} \tag{2}$$

Also, we have the kinematic conditions

$$\frac{dr}{dt} = \frac{\partial\phi}{\partial r}; \quad \frac{dz}{dt} = \frac{\partial\phi}{\partial z}. \tag{3}$$

Given the position of the interface and the value of p there at any instant, if the Laplacian problem for ϕ can be solved $\underset{\sim}{v}$ may be computed. The new position of the interface an instant later can now be found and the new value of ϕ on the interface from (2). The calculation of $\underset{\sim}{v} = \nabla\phi$ is then repeated at this later time.

Although mathematically the boundary value problem for ϕ is simple the changing shape of the contour in general makes it impossible to obtain analytical solutions. Plesset and Chapman [2] found the potential with a wall present by setting up a number of meshes based on cylindrical polar coordinates, with the mesh length increasing with distance from the cavity. They satisfied the boundary condition at infinity by fitting the potential and its spherical derivative to a Legendre coefficient expansion using an iterative method which converged quickly as long as the outer mesh boundary was far from the cavity. Though no calculation times are quoted, this may not be a particularly fast method of finding the potential. Since the calculation has to be repeated at each time step we have tried another method which is very fast but turns out to be only partially successful. A somewhat slower version of the same method which in theory should not present the difficulties experienced with the first version is at present under development.

The cavity is represented by a number of points along its contour. At each time step, once the potential has been found, the velocity is calculated and the contour moved to its new position within the framework of a contour moving code developed at Culham by Hughes, Roberts and Zabusky. The changing shape of a typical collapsing bubble up to the time that the jet starts to form can be seen in Fig. 1. Three variables are defined at each point of the contour: its position (r,z) and the value of the potential ϕ. The values of r, z and ϕ are updated at each time step using a leapfrog scheme in time with matching to avoid numerical instability according to (2) and (3) where the right-hand side of these equations is determined from $\nabla\phi$. A typical run lasts a hundred time steps though with experience this figure may be reduced. The details of this part of the scheme are not discussed here, only the calculation of the potential at each time step. This is done in a subroutine incorporated into the code which already provides graphical output and other administrative routines.

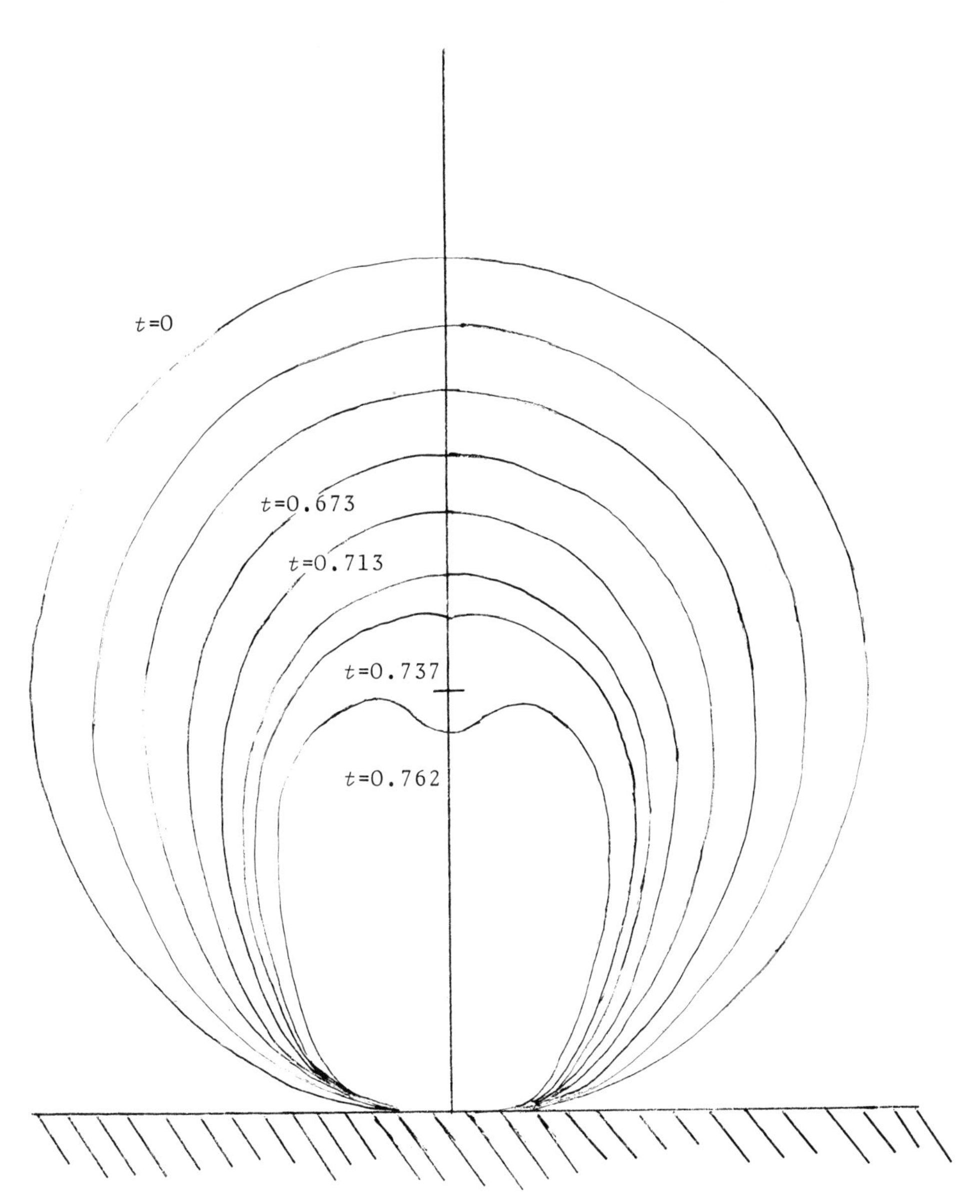

Fig. 1. Bubble collapse onto a rigid wall. There are 11 sources on the axis and initially 11 points on the half-contour, rising to 21 at time t=0.762. The maximum normalised least mean square deviation in the potential rises from zero to 1.9%. Investigation of the individual errors at each point on the contour shows they oscillate in sign almost alternately from point to point. The errors are lowest on the sides of the bubble, moderate at the wall and highest around the jet.

It is known that solutions of Laplace's equation can be represented by suitable distributions of sources, sinks and vortex rings, etc. All such solutions determined by boundary conditions on the cavity surface can be represented internally or externally by a source distribution distributed over the cavity surface. If $G(\underset{\sim}{r},s')$ is the potential at position $\underset{\sim}{r}$ due to a source density $\rho(s')$, where s' refers in this axisymmetric situation to the distance along the contour, we may write

$$\phi(\underset{\sim}{r}) = \int \rho(s')G(\underset{\sim}{r},s')ds' \qquad (4)$$

and

$$\nabla\phi(\underset{\sim}{r}) = \int \rho(s')\nabla_{\underset{\sim}{r}}G(\underset{\sim}{r},s')ds'. \qquad (5)$$

$\rho(s')$ is at present an unknown distribution. In the presence of a flat wall G includes the potential due to the image distribution. If ϕ is given on the contour (4) may be written there

$$\phi(s) = \int \rho(s')G(s,s')ds'. \qquad (6)$$

If this integral equation for ρ can be inverted, $\nabla\phi$ may be found from (5) everywhere and in particular on the contour. If N unknown values of ρ are assigned at N points on the contour (or half contour given the axisymmetry) at which the N values of ϕ are given, the integral equation (6) may be written as a set of N simultaneous equations for the N values of ρ given the N values of ϕ.

In matrix form we may write

$$\underline{\underline{G}}^{N\times N}\underset{\sim}{\rho}^{N} = \underset{\sim}{\phi}^{N} . \qquad (7)$$

Some care is necessary in the formulation of the $N\times N$ matrix $\underline{\underline{G}}$ since the Green's function G in the axisymmetric case has a logarithmic singularity ($\sim\log|s-s'|$) as $s \to s'$. This method is under development at present.

The number of points on the (half) contour N may range perhaps from 10 to 50, depending on the definition required and the contour shape. The resulting $N\times N$ matrix is much smaller than the matrix resulting from a finite difference formulation of the problem, which would $\sim N^2 \times N^2$. It is not sparse but its inversion time is small. Its elements involve elliptic functions.

We have tried a somewhat different approach, with only partial success, to reduce the time for matrix inversion and calculation. Instead of representing the potential by an unknown distribution of sources placed on the contour, the sources are placed on the axis. Though a source distribution on the axis yields an axisymmetric solution of Laplace's equation, it is not clear that all axisymmetric solutions can be represented in this way. It is hoped that the potentials occurring in bubble collapse problems might admit such a representation, though as jet formation proceeded and the axis available for the charge distribution contracted the accuracy of the representation would be expected to decrease.

The potential ϕ is specified at N points on the contour and the unknown source ρ assigned at M points on the axis, where $M \leqslant N$. The matrix $\underline{\underline{G}}$ now consists of elements involving the distance between the various points rather than elliptic integrals, and since none of the points need coincide singularities are avoided. If $M = N$ the resulting $N \times N$ simultaneous equations are solved for ρ as before. However this does not reduce the number of equations and gives no indication of the accuracy of the representation of ϕ by the source distribution. If $M < N$, the source distribution cannot, in general, represent the potential exactly but a set of values of ρ which give the least mean square deviation of ϕ from the prescribed values can be found and requires the inversion of an $M \times M$ matrix. Explicitly the equations for $\underset{\sim}{\rho}^M$ are

$$(\underline{\underline{G}}^T \underline{\underline{G}})\underset{\sim}{\rho} = \underline{\underline{G}}^T \underset{\sim}{\phi}. \tag{8}$$

The resulting least mean square error or maximum error, suitably normalised, is some indication of the accuracy of the process.

For a circular contour the integral equation relating the source distribution on the axis to ϕ on the contour has an analytic inversion formula, provided the boundary values of ϕ are continuously differentiable (*e.g.* [1] p.939). Using this as a test we found that relatively few sources on the axis, evenly spread over about half the axis, provided remarkably good results, and that in general no source was required at the ends of the axis, perhaps because the condition $\frac{\partial \phi}{\partial r} = 0$ requires zero source density there. Typical results for $\phi = 1$ or $\phi = \cos\theta$ specified at $N = 20$ to 25 points on the half contour gave fractional maximum errors in ϕ ranging from 10^{-3} to 10^{-5} as M varied from 6 to 10, with errors in $\nabla\phi$ some 10 times larger. A more precise comparison shows that the errors depend on the source spacing adopted and were smaller in these test cases where the sources were closely packed around the centre of the axis. For $N >> M$ the errors appeared to reach a limiting value depending on M as N increased. This value quantifies in some sense the ability of a small number of sources on the axis to fit a smoothly varying potential specified on the contour.

When this method was incorporated into the contour moving programme the results were only partially successful. In most cases the sources were spaced evenly over a variable fraction of the available axis. It was found that even in the early stages of the calculation, when the cavity was still approximately spherical, the accuracy was low when the number of sources on the axis was less than the number of points on the contour (typically a maximum error of a few per cent when the number of sources was 7 and the number of points on the half contour 21). The contour of the cavity also showed signs of ripples forming at wavelengths based on the spacing of the points defining the contour, despite the fact that the interface of a spherically symmetric bubble collapsing under a constant pressure difference is only weakly unstable, in the sense that small perturbations of the interface are oscillatory with an amplitude that increases rapidly only at the end of the collapse [3].

This problem appeared to be alleviated in the early stages by choosing the same number of sources on the axis and points on the half-contour. The least mean square fitting (8) now yields the same answer as the direct inversion of (7) and the zero error in the fitting process does not reflect the accuracy with which the source distribution can be fitted to the potential distribution. As the contour is deformed the program automatically includes extra points where necessary to maintain its definition and since the number of sources on the axis cannot be increased indefinitely the least mean square fitting process is invoked in the later stages of collapse. The error rises progressively reaching a few per cent as the jet starts to form. For the collapse shown in Fig. 1 the number of sources on the axis was fixed at 11, and the number of points defining the half contour rose from 11 to 21, by which time the error was 1.9%.

The conclusion seems to be that on contours of this shape the potentials that occurred could not be adequately represented by a source distribution on the axis. This is disappointing, though it had been expected that if the representation did become inaccurate it would do so at this stage. On the basis of our limited tests we can offer no conclusive reasons why the representation is so poor in the early stages.

To improve the accuracy a number of off-axis singularities were then added, in the form of vortex rings positioned somewhere in the cavity. It is more difficult to decide where to place these rings than it is to decide on the source position on the axis, and the results obtained showed little increase in accuracy. Refinements in the positioning of the vortex rings might have improved the accuracy, but we felt that even if it did, automating the positioning in a programme for use by a non-specialist might be difficult. We therefore decided to revert to the method involving a source distribution on the contour which, although rather slower, has the merit of being known to work in principle.

Though the method of solving Laplace's equation described above is not sufficiently accurate for the later stages of jet formation in a collapsing bubble, it takes a negligible amount of computer time. It may yet be possible to use it to describe the earlier stages and the slower but theoretically sounder method at present under development for the later stages.

REFERENCES

1. Morse, P.M. and Feshbach, H., "Methods of theoretical physics," McGraw Hill (1953).
2. Plesset, M.S. and Chapman, R.B., *J. Fluid Mech.* **47**, 2, 283-290 (1971).
3. Plesset, M.S. and Mitchell, T.P., *Q. Appl. Math.* **13**, 419 (1956).

Author Index

Subject Index